# Clinical Liquid Chromatography

Volume I

Analysis of Exogenous Compounds

Editors

**Pokar M. Kabra**
and
**Laurence J. Marton**

Department of Laboratory Medicine
School of Medicine
University of California, San Francisco
San Francisco, California

CRC Press, Inc.
Boca Raton, Florida

**Library of Congress Cataloging in Publication Data**

Main entry under title:

Clinical liquid chromatography.

Bibliography: p.
Includes index.
Contents: v. 1. Analysis of exogenous compounds.
v. 2. Analysis of endogenous compounds.
1. Liquid chromatography. 2. Chemistry, Clinical—
Technique. I. Kabra, Pokar, M. II. Marton, Laurence J.
[DNLM: 1. Chromatography, Liquid. QD 79.C454 C641]
QP519.9.L55C54 1984      616.07'56      83-3886
ISBN-0-8493-6637-2 (v. 1)
ISBN 0-8493-6638-0 (v. 2)

This book represents information obtained from authentic and highly regarded sources. Reprinted material is quoted with permission, and sources are indicated. A wide variety of references are listed. Every reasonable effort has been made to give reliable data and information, but the author and the publisher cannot assume responsibility for the validity of all materials or for the consequences of their use.

All rights reserved. This book, or any parts thereof, may not be reproduced in any form without written consent from the publisher.

Direct all inquiries to CRC Press, Inc., 2000 Corporate Blvd., N.W., Boca Raton, Florida, 33431.

© 1984 by CRC Press, Inc.

International Standard Book Number 0-8493-6637-2 (Volume I)
International Standard Book Number 0-8493-6638-0 (Volume II)

Library of Congress Card Number 83-3886
Printed in the United States

# PREFACE

Liquid chromatography (LC) is widely used for the analysis of exogenous and endogenous constituents in physiological fluids. These volumes are designed to serve as a reference for scientists involved in applying these techniques to clinical and basic science use and are a compilation of a variety of accepted and tested liquid chromatographic methods. In most instances, the authors review the basic concepts underlying the development of the particular LC method and the present specific details of each procedure. All the chapters use a similar format in describing methods which are based upon the cumulative experience of individuals who have been pioneers in the field. These methods should yield reproducible results when carried out exactly as described in the text. Subsequently, some may wish to modify these methods to suit their own needs; however, this should be done only after careful evaluation of the different analytical parameters described in the text. One important feature of these volumes is that alternate methods of analysis are described for many important compounds. By utilizing these procedures, laboratories should be able to provide clinicians with objective laboratory data. In addition, many of the methods described can serve as reference methods to evaluate the accuracy of a variety of nonchromatographic techniques. Every effort was made to include most clinically important endogenous and exogenous constituents. However, the tremendous growth of LC and the constraints of space has resulted in the omission of some compounds.

We sincerely thank the contributors to this volume for their efforts and our families for their patience.

## THE EDITORS

**Dr. Pokar M. Kabra** is Associate Professor of Laboratory Medicine at the University of California, San Francisco. Dr. Kabra received his B.S. (Chemistry) and B.S. (Pharmacy) from the University of Bombay, Bombay, India, and his Ph.D. in Medicinal Chemistry from the University of Kansas, Lawrence, Kansas. Dr. Kabra is a member of five professional and/or scientific organizations and has published more than 50 research papers including 20 reviews and monographs. He has also coedited a book on the subject of liquid chromatography in clinical analysis. His current research interest includes the development of very high speed liquid chromatography in therapeutic drug monitoring, analysis of endogenous constituents, and development of unique precolumn derivatization techniques for clinical analysis.

**Laurence J. Marton** is Professor of Laboratory Medicine and Neurological Surgery and Chairman of the Department of Laboratory Medicine at the University of California at San Francisco, School of Medicine. He is also a member of the Brain Tumor Research Center. He received the M.D. degree (1969) from the Albert Einstein College of Medicine, Bronx, New York. He has served as an advisor to the National Institutes of Health and the American Cancer Society, and was a recipient of a National Cancer Institute Research Career Development award. His research has focused on the cell and molecular biology of polyamines as they relate to the diagnosis, monitoring, and therapy of cancer and on the application of liquid chromatography to the clinical laboratory.

# CONTRIBUTORS

**Laura A. Allison, M.S.**
Research Chemist
Bioanalytical Systems, Inc.
West Lafayette, Indiana

**Norbert Blanckaert, M.D., Ph.D.**
Lecturer
Department of Medical Research
Catholic University of Leuven
Research Associate of the National
  Fund of Scientific Research
Leuven, Belgium

**Craig S. Bruntlett, Ph.D.**
Director of New Product Development
Bioanalytical Systems, Inc.
West Lafayette, Indiana

**Mei-Ling Chen, Ph.D.**
Research Associate
Department of Pharmacodynamics
College of Pharmacy
University of Illinois at Chicago
Chicago, Illinois

**Win L. Chiou, Ph.D.**
Professor of Pharmacodynamics
Department of Pharmacodynamics
College of Pharmacy
University of Illinois at Chicago
Chicago, Illinois

**Thomas P. Davis, Ph.D.**
Department of Pharmacology
University of Arizona College of
  Medicine
Tucson, Arizona

**Charles W. Gehrke, Ph.D**
Professor of Biochemistry and Manager
  of Experiment Station Chemical
  Laboratories
University of Missouri
Columbia, Missouri

**Charles W. Gehrke, Jr.***

**Klaus O. Gerhardt, Ph.D**
Department of Biochemistry
Experiment Station Chemical Laboratories
University of Missouri
Columbia, Missouri

**Thomas J. Good**
Vice President, Research
Analytichem International
Harbor City, California

**George R. Gotelli**
Chemistry Specialist
Department of Laboratory Medicine
University of California, San Francisco
San Francisco, California

**T. W. Guentert, Ph.D**
Lecturer in Pharmacy
School of Pharmacy
University of Basel and
F. Hoffmann-La Roche & Co.
Basel, Switzerland

**Karen K. Haak**
Methods Development Chemist
Dionex Corporation
Sunnyvale, California

**Martha R. Harkey, Ph.D.**
Director Pharmacokinetics Service
French Hospital Medical Center
San Francisco, California

**Edward L. Johnson**
Director of Technology
Dionex Corporation
Sunnyvale, California

**Pokar M. Kabra, Ph.D.**
Associate Professor
Department of Laboratory Medicine
School of Medicine
University of California, San Francisco
San Francisco, California

\* Deceased

**Peter T. Kissinger, Ph.D**
Department of Chemistry
Purdue University
West Lafayette, Indiana

**Kristi J. Klippel**
Applications Chemist
Bioanalytical Systems, Inc.
West Lafayette, Indiana

**David D. Koch, Ph.D**
Department of Pathology
 and Laboratory Medicine
University of Wisconsin Hospital and
 Clinics
Madison, Wisconsin

**Kenneth C. Kuo, M.S.**
Senior Research Chemist
University of Missouri
Columbia, Missouri

**J. J. Lauff**
Research Associate
Research Laboratories
Eastman Kodak Company
Rochester, New York

**Ellen M. Levin, M.S.**
Medical Technologist
Laboratory Medicine and Brain
 Tumor Research Center
University of California, San Francisco
San Francisco, California

**Victor A. Levin, M.D.**
Professor
Departments of Neurological Surgery,
 Pharmacology, and Pharmaceutical
 Chemistry
Associate Director
Brain Tumor Research Center
University of California, San Francisco
San Francisco, California

**Emil T. Lin**
Assistant Professor
School of Pharmacy
University of California, San Francisco
San Francisco, California

**Warren P. Lubich**
Brain Tumor Research Center
Department of Neurological Surgery
School of Medicine
University of California, San Francisco
San Francisco, California

**Laurence J. Marton, M.D.**
Professor of Laboratory Medicine and
 Neurological Surgery
Chairman
Department of Laboratory Medicine
School of Medicine
University of California, San Francisco
San Francisco, California

**Ginny S. Mayer**
Applications Chemist
Bioanalytical Systems, Inc.
West Lafayette, Indiana

**Julie Morris**
Applications Chemist
Bioanalytical Systems, Inc.
West Lafayette, Indiana

**William E. Rich**
Vice President
Dionex Corporation
Sunnyvale, California

**Thomas G. Rosano, Ph.D.**
Associate Director of Clinical Chemistry
Associate Professor of Biochemistry
Albany Medical Center
Albany, New York

**Timothy D. Schlabach**
Senior Chemist
Varian Instrument Group
Walnut Creek, California

**Gary J. Schmidt**
Research Chemist
Perkin-Elmer Corporation
Applied Research Group
Norwalk, Connecticut

**Nikolaus Seiler, Ph.D.**
Associate Scientist
Centre de Recherche
Merrell International
Filiale de Dow Chemical Co.
Strasbourg, France

**Wayne L. Settle, Ph.D.**
Assistant Director
Pharmacokinetics Service
French Hospital Medical Center
San Francisco, California

**Ronald E. Shoup, Ph.D.**
Vice President for Research
Bioanalytical Systems, Inc.
West Lafayette, Indiana

**J. Stuart Soeldner, M.D.**
Associate Professor of Medicine
Brigham and Women's Hospital
Harvard Medical School
Senior Investigator
Elliott P. Joslin Research Laboratory
Joslin Diabetes Center, Inc.
Boston, Massachusetts

**Mark Sothmann, Ph.D**
Assistant Professor of Exercise
  Physiology
Department of Human Kinetics
University of Wisconsin
Milwaukee, Wisconsin

**Harold S. Starkman, M.D.**
Research Fellow in Medicine
Brigham and Women's Hospital
Harvard Medical School
Elliott P. Joslin Research Laboratory
Joslin Diabetes Center, Inc.
Boston, Massachusetts

**James T. Taylor, B.S.**
Chief Chemist and Department Supervisor
Special Chemistry
Pathology Laboratories, Ltd.
Hattiesburg, Mississippi

**M. Michael Thaler, M.D.**
Professor
Department of Pediatrics
University of California, San Francisco
San Francisco, California

**James G. Vaughn**
Product Manager
Research Chromatography Products
LKB Instruments, Inc.
Gaithersburg, Maryland

**Marilyn Tuttleman Wacks, B.S.M.T.
 (ASCP)**
Senior Technician
Elliott P. Joslin Research Laboratory
Joslin Diabetes Center, Inc.
Boston, Massachusetts

**Jeffrey H. Wall**
Senior Clinical Laboratory Technologist
  Specialist
Department of Laboratory Medicine
University of California, San Francisco
San Francisco, California

**Joanne O. Whitney, Ph.D.**
Adjunct Associate Professor
Department of Pharmaceutical Chemistry
University of California, San Francisco
San Francisco, California

**Charles H. Williams, Ph.D**
Departments of Anesthesiology and
  Biochemistry
Texas Tech Regional Medical Center
El Paso, Texas

**T. W. Wu**
Research Associate
Research Laboratories
Eastman Kodak Company
Rochester, New York

**Lane S. Yago**
Manager
Marketing Services
Analytichem International
Harbor City, California

**Robert W. Zumwalt, Ph.D**
Research Associate
Department of Biochemistry
University of Missouri
Columbia, Missouri

# TABLE OF CONTENTS

## Volume I

Chapter 1
Acetaminophen and Phenacetin by Ultraviolet Detection .................................. 1
**George R. Gotelli and Jeffrey H. Wall**

Chapter 2
Determination of Acetaminophen in Plasma by Liquid Chromatography/
Electrochemistry.......................................................................................... 5
**Kristi Klippel**

Chapter 3
Determination of ϵ-Aminocaproic Acid...................................................... 11
**Gary J. Schmidt**

Chapter 4
Analysis of Aminoglycoside Antibiotics by Precolumn Fluorescence Derivatization..... 17
**Wayne L. Settle and Martha R. Harkey**

Chapter 5
Serum Amikacin with Spectrophotometric Detection ..................................... 25
**Pokar M. Kabra**

Chapter 6
Serum Gentamicin with Spectrophotometric Detection .................................. 29
**Pokar M. Kabra**

Chapter 7
Serum Tobramycin with Spectrophotometric Detection................................. 33
**Pokar M. Kabra**

Chapter 8
Antidysrhythmic Drugs by Ultraviolet/Fluorescence Detection .......................... 39
**Pokar M. Kabra**

Chapter 9
Procainamide and *N*-Acetyl Procainamide by Ultraviolet Detection ..................... 47
**George R. Gotelli and Jeffrey H. Wall**

Chapter 10
Propranolol by Fluorescence Detection ...................................................... 53
**George R. Gotelli and Jeffrey H. Wall**

Chapter 11
Quinidine by Fluorescence Detection......................................................... 57
**George R. Gotelli and Jeffrey H. Wall**

Chapter 12
Quinidine and Its Metabolites by Fluorescence Detection ............................... 63
**Theodore W. Guentert**

Chapter 13
Simultaneous Analysis of Carbamazepine, Ethosuximide, Phenobarbital, Phenytoin, and Primidone .................................................................................. 71
**Pokar M. Kabra**

Chapter 14
Simultaneous Very High Speed Liquid-Chromatographic Analysis of Ethosuximide, Primidone, Phenobarbital, Phenytoin, and Carbamazepine in Serum ........................ 77
**Pokar M. Kabra**

Chapter 15
Simultaneous Analysis of Common Sedatives in Serum ................................. 83
**Pokar M. Kabra**

Chapter 16
Pentobarbital by Ultraviolet Detection ................................................ 89
**Jeffrey H. Wall and George R. Gotelli**

Chapter 17
Chloramphenicol by Ultraviolet Detection ............................................. 93
**George R. Gotelli and Jeffrey H. Wall**

Chapter 18
Chlorpromazine in Plasma by Liquid Chromatography/Electrochemistry ................ 97
**Julie Morris and Ronald E. Shoup**

Chapter 19
Oxazepam, Diazepam, and $N$-Desmethyldiazepam in Human Blood by Ultraviolet Detection .................................................................................. 101
**Pokar M. Kabra**

Chapter 20
Chlorthalidone in Blood and Urine .................................................... 107
**Emil T. Lin**

Chapter 21
Furosemide in Plasma and Urine ...................................................... 111
**Emil T. Lin**

Chapter 22
Hydrochlorothiazide in Plasma and Urine ............................................. 115
**Emil T. Lin**

Chapter 23
Methyclothiazide in Blood and Urine ................................................. 119
**Emil T. Lin**

Chapter 24
Triamterene and Its Metabolite in Plasma and Urine .................................. 123
**Emil T. Lin**

Chapter 25
Exogenous Glucocorticoids, Prednisone, and Prednisolone in Plasma .................. 129
**Emil T. Lin**

Chapter 26
5-Fluorocytosine by Ultraviolet Detection .............................................. 135
**George R. Gotelli and Jeffrey H. Wall**

Chapter 27
Simultaneous Determination of Methotrexate and Its Metabolites in Plasma, Saliva, and Urine ..................................................................................... 139
**Mei-Ling Chen and Win L. Chiou**

Chapter 28
Analysis of Misonidazole and Desmethylmisonidazole ................................. 147
**Ellen M. Levin and Victor A. Levin**

Chapter 29
Morphine in Plasma by Liquid Chromatography ....................................... 153
**Kristi Klippel**

Chapter 30
Theophylline by Ultraviolet Detection ................................................... 159
**George R. Gotelli and Jeffrey H. Wall**

Chapter 31
Very High Speed Liquid Chromatographic Analysis of Theophylline in Serum ........ 163
**Pokar M. Kabra**

Chapter 32
Simultaneous Reversed-Phase Liquid Chromatographic Analysis of Amitriptyline, Nortriptyline, Imipramine, Desipramine, Doxepin, and Nordoxepin ......................... 167
**Pokar M. Kabra**

Chapter 33
Determination of Amitriptyline, Imipramine, Nortriptyline, and Desipramine .......... 173
**Gary J. Schmidt**

Chapter 34
Imipramine, Desipramine, and Metabolites by Liquid Chromatography/Electrochemistry ........................................................................... 179
**Julie Morris and Ronald E. Shoup**

Chapter 35
Determination of Thiols by Liquid Chromatography/Electrochemistry .................. 185
**Laura A. Allison and Ronald E. Shoup**

Chapter 36
Screening Toxic Drugs in Serum with Gradient Liquid Chromatography ............... 191
**Pokar M. Kabra**

Chapter 37
Sample Preparation for Liquid Chromatographic Analysis ............................. 197
**Lane S. Yago and Thomas J. Good**

Index ............................................................................................. 209

## TABLE OF CONTENTS

### Volume II

Chapter 1
Amino Acid Analysis of Physiological Fluids by Ion-Exchange Chromatography ........ 1
**James G. Vaughn**

Chapter 2
Determination of Phenylalanine ...................................................... 11
**Gary J. Schmidt**

Chapter 3
Analysis of Conjugated Bile Salts in Bile and Duodenal Aspirates by Radial Compression HPLC ............................................................................. 17
**J. O. Whitney and M. M. Thaler**

Chapter 4
Determination of Bilirubin and Its Ester Conjugates .................................. 23
**Norbert Blanckaert**

Chapter 5
Determination of Bilirubin Species in Serum Including Strongly Protein-Linked Bilirubin ($\delta$) ........................................................................... 31
**J. J. Lauff and T. W. Wu**

Chapter 6
Urinary Catecholamines by Liquid Chromatography/Electrochemistry .................. 41
**Laura A. Allison and Ronald E. Shoup**

Chapter 7
Determination of Plasma Norepinephrine by Liquid Chromatography/Electrochemistry 47
**Ronald E. Shoup and Ginny S. Mayer**

Chapter 8
Precolumn Derivatization, HPLC, and Fluorescence Measurement of Biogenic Amines in Biological Materials ............................................................. 53
**Thomas P. Davis, Charles W. Gehrke, Jr., Klaus O. Gerhardt, Charles H. Williams, and Charles W. Gehrke**

Chapter 9
Determination of Urinary Normetanephrine, Metanephrine, and 3-Methoxytyramine by Liquid Chromatography/Electrochemistry ................................................ 65
**Ronald E. Shoup and Peter T. Kissinger**

Chapter 10
Determination of Urinary Vanilmandelic Acid (VMA) by Liquid Chromatography/
Electrochemistry.................................................................71
**Mark S. Sothman and Craig S. Bruntlett**

Chapter 11
Determination of Urinary Homovanillic Acid (HVA) by Liquid Chromatography/
Electrochemistry.................................................................77
**Mark S. Sothman and Craig S. Bruntlett**

Chapter 12
Urinary Homovanillic Acid by Fluorescent Detection ...................................83
**Thomas G. Rosano**

Chapter 13
Determination of Urinary 3-Methoxy-4-Hydroxyphenylglycol (MHPG) by Liquid Chromatography/Electrochemistry .......................................................87
**Craig S. Bruntlett and Mark S. Sothman**

Chapter 14
MHPG (3-Methoxy-4-Hydroxyphenylethylene Glycol) in Urine by Fluorescence
Detection .......................................................................95
**James T. Taylor**

Chapter 15
Serum Cortisol with Ultraviolet Detection................................................103
**Pokar M. Kabra**

Chapter 16
Cortisol by Fluorescence Detection......................................................107
**George R. Gotelli and Jeffrey H. Wall**

Chapter 17
Placental Estriol in Urine by Ultraviolet Detection ......................................113
**George R. Gotelli and Jeffrey H. Wall**

Chapter 18
Estriol in Pregnancy Urine by Fluorescence Detection ...................................117
**James T. Taylor**

Chapter 19
Determination of Urinary Estriol by Liquid Chromatography/Electrochemistry .........123
**Julie Morris and Ronald E. Shoup**

Chapter 20
Unconjugated Serum Estriol with Fluorometric Detection ................................127
**Pokar M. Kabra**

Chapter 21
Hemoglobin $A_{1c}$ by High Pressure Liquid Chromatography (HPLC) ....................131
**Marilyn Tuttleman Wacks, Harold S. Starkman, and J. Stuart Soeldner**

**Chapter 22**
Analysis of Ribonucleosides in Biological Matrices ................................... 139
**Charles W. Gehrke, Robert W. Zumwalt, and Kenneth C. Kuo**

**Chapter 23**
The Determination of Organic Acids in Physiological Fluids by Ion Chromatography: Plasma Pyruvate and Lactate and Urinary Oxalate ................................... 155
**Karen K. Haak, William E. Rich, and Edward Johnson**

**Chapter 24**
Erythrocyte Protoporphyrin, Zinc Protoporphyrin, and Coproporphyrin by Fluorescence Detection ................................... 167
**George R. Gotelli and Jeffrey H. Wall**

**Chapter 25**
Analysis of Putrescine, Spermidine, and Spermine in Biological Fluids with a Modified Amino Acid Analyzer ................................... 173
**Laurence J. Marton and Warren P. Lubich**

**Chapter 26**
Separation of Ion Pairs of Polyamines and Related Compounds on Reversed Phases ... 179
**Nikolaus Seiler**

**Chapter 27**
Separation and On-Line Quantitation of Lactate Dehydrogenase and Creatine Kinase Isoenzymes in Human Sera by High Performance Liquid Chromatography ................. 189
**Timothy D. Schlabach**

**Chapter 28**
Vitamin $D_3$ and its Physiologically Active Metabolites ................................... 199
**J. O. Whitney and M. M. Thaler**

**Chapter 29**
Urinary 5-Hydroxy-3-Indoleacetic Acid by Fluorescent Detection ..................... 205
**Thomas G. Rosano**

**Chapter 30**
Urinary Tryptophan, Serotonin, and 5-Hydroxyindole Acetic Acid by Liquid Chromatography/Electrochemistry ................................... 209
**David D. Koch and Peter T. Kissinger**

**Chapter 31**
Determination of Serotonin in Serum by Liquid Chromatography with Precolumn Sample Enrichment and Electrochemical Detection ................................... 217
**David D. Koch and Peter T. Kissinger**

Index ................................... 223

Chapter 1

# ACETAMINOPHEN AND PHENACETIN BY ULTRAVIOLET DETECTION

**George R. Gotelli and Jeffrey H. Wall**

## TABLE OF CONTENTS

I. Introduction ................................................................. 2

II. Principle .................................................................... 2

III. Materials and Method ........................................................ 2
    A. Equipment ............................................................. 2
    B. Reagents .............................................................. 2
    C. Standards ............................................................. 2
    D. Procedure ............................................................. 2
    E. Calculations .......................................................... 3

IV. Results ..................................................................... 3
    A. Optimization of Chromatography ........................................ 3
    B. Linearity ............................................................. 3
    C. Recovery .............................................................. 4
    D. Reproducibility ....................................................... 4
    E. Accuracy .............................................................. 4
    F. Sensitivity ........................................................... 4
    G. Interferences ......................................................... 4

V. Comments .................................................................... 4

References ...................................................................... 4

## I. INTRODUCTION

Acetaminophen and phenacetin are commonly used aspirin alternatives and are usually well-tolerated at the recommended dose. Large doses, however, may be toxic and have been associated with lethal hepatic necrosis and renal failure. The available colorimetric methods for acetaminophen are nonspecific and frequently insensitive. In contrast, liquid chromatography offers specificity and excellent sensitivity[1] (Figure 1).

## II. PRINCIPLE

Acetaminophen, phenacetin, and added internal standard are extracted by ethyl acetate from buffered serum. After centrifugation, the ethyl acetate is decanted, evaporated, and the residue is redissolved in methanol and injected onto the liquid chromatographic column. The drugs are eluted from the column with acetonitrile-phosphate buffer at 50°C, detected by their absorbance at 254 nm, and quantitated by peak height ratios.

## III. MATERIALS AND METHOD

### A. Equipment
A liquid chromatography (LC) system equivalent to the following is recommended. A model 601 (Perkin-Elmer Corp., Norwalk, Conn.) with a Rheodyne® 7105 valve (Rheodyne, Cotati, Calif.), a 1-MV recorder, a temperature controlled oven (Perkin-Elmer, Model LC100) set at 50°C, a reverse-phase octadecylsilane column similar to a Waters® C 18 µBondapak® 30 cm × 4 mm i.d. (Waters Associates, Inc., Milford, Mass.) or a Whatman® ODS 3, 25 cm × 4.6 mm i.d. (Whatman, Inc., Clifton, N.J.) and a detector capable of monitoring at 254 nm.

### B. Reagents
Acetonitrile, uv grade (Burdick and Jackson Laboratories Inc., Muskegon, Mich.). The phosphate buffer (pH 4.4) is prepared by adding 300 µℓ of 1 mol/ℓ potassium dihydrogen phosphate and 50 µℓ of 4.4 mol/ℓ phosphoric acid to 1800 mℓ of distilled water. The mobile phase consists of 19 parts of acetonitrile and 81 parts (by volume) of phosphate buffer (ph 4.4).

### C. Standards
**Drug standard** — Dissolve acetaminophen (200 mg), acetoacetanilide (400 mg), and phenacetin (200 mg) in a final volume of 1 ℓ of methanol.
**Internal standard** — A stock solution is prepared by dissolving 500 mg of acetoacetanilide in a final volume of 1 ℓ of methanol. A working solution is prepared daily by diluting the stock internal standard tenfold in water. All drugs were purchased from Sigma Chemical Co., St. Louis. Quality control serum can be prepared by dissolving an appropriate amount of acetaminophen and phenacetin in 5 to 10 mℓ of methanol and then diluting to a final volume with water. A measured amount of the drug solution is then added to serum, mixed, and frozen in aliquots. These controls are stable for at least 3 months when frozen.

### D. Procedure
Transfer 0.5 mℓ of serum, 0.5 mℓ of working internal standard, and 0.5 mℓ of 1 mol/ℓ phosphate buffer (pH 7.0) to a 12 mℓ stoppered tube. Add 7mℓ of ethyl acetate and shake for 10 min on a mechanical shaker. Centrifuge 5 min, decant the ethyl acetate, and evaporate it at 70°C under reduced pressure. Dissolve the residue in 50 µℓ of methanol and inject 10 to 20 µℓ onto the column at a flow rate of 3 mℓ/min at 50°C, with the detector set at 254 nm.

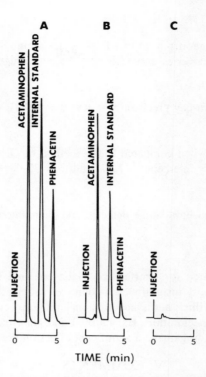

FIGURE 1. (A) Drug reference standard; (B) serum containing 32 mg of acetaminophen and 3 mg of phenacetin per liter; (C) drug-free serum.

## E. Calculations

Calculate the response factor of each drug relative to the internal standard following direct injection of the drug standard as follows:

$$\text{Response factor (RF)} = \frac{\text{peak height of Int. Std.}}{\text{peak height of drug} \times 2}$$

Calculate the concentration of the unknown drug as follows:

$$\frac{\text{peak height of drug}}{\text{peak height of Int. Std.}} \times \text{RF of respective drug} \times 50 = \mu g/m\ell \text{ unknown}$$

Alternatively, serum-based standards, containing known amounts of acetaminophen and phenacetin, can be processed and used to prepare a standard curve.

## IV. RESULTS

### A. Optimization of Chromatography

A pH of 4.4 was determined to be optimal for resolution of these drugs. The elevated temperature used during chromatography reduces the viscosity of the mobile phase and improves the resolution of the compounds.

### B. Linearity

This procedure is linear to 100 mg/$\ell$ for acetaminophen and 40 mg/$\ell$ for phenacetin.

## C. Recovery

Analytical recoveries ranged from 94 to 109% for acetaminophen and from 90 to 110% for phenacetin. Absolute recoveries exceeded 80% for both drugs.

## D. Reproducibility

The day-to-day and within-day precision was less than 6% (CV) for both drugs.

## E. Accuracy

The regression analyses when compared to the method of Glynn and Kendal[2] was $r = 0.989$, slope $= 0.996$, the y-intercept $= 7.17$, and $n = 30$.

## F. Sensitivity

Acetaminophen and phenacetin can be detected and quantitated at a concentration of 0.5 mg/$\ell$ of blood.

## G. Interferences

There were 33 drugs tested for interference. Salicylate and theophylline coelute with acetaminophen, however, only insignificant amounts of salicylate are extracted by the described procedure. Theophylline, at therapeutic concentrations, can potentially increase the apparent acetaminophen concentration. Butabarbital and phensuximide coelute with phenacetin, but neither of these drugs have a significant effect on the phenacetin value.

## V. COMMENTS

This method provides the necessary sensitivity to quantitate therapeutic blood concentration and the linearity to quantitate toxic concentrations. Sample processing is rapid and requires as little as 0.1 m$\ell$ of serum. Extraction by ethyl acetate at pH 7.0 was found to be necessary in order to eliminate potential interference by endogenous compounds as well as to prevent interference from large amounts of salicylates, commonly used in combination with acetaminophen and phenacetin.

The accepted therapeutic ranges for acetaminophen and phenacetin are 5 to 15 and 0.1 to 2.0 $\mu$g/m$\ell$, respectively. Blood concentrations of acetaminophen $>200$ $\mu$g/m$\ell$ 4 hr after ingestion, or $>50$ $\mu$g/m$\ell$, 12 hr after ingestion have been associated with hepatic toxicity.

## REFERENCES

1. **Gotelli, G. R., Kabra, P. M., and Marton, L. J.,** Determination of acetaminophen and phenacetin in plasma by high-pressure liquid chromatography, *Clin. Chem.*, 23, 957, 1977.
2. **Glynn, J. P. and Kendal, S. E.,** Paracetamol measurement, *Lancet*, i, 1147, 1975.

Chapter 2

# DETERMINATION OF ACETAMINOPHEN IN PLASMA BY LIQUID CHROMATOGRAPHY/ELECTROCHEMISTRY

**Kristi Klippel**

## TABLE OF CONTENTS

| | | |
|---|---|---|
| I. | Introduction | 6 |
| II. | Principle | 6 |
| III. | Materials and Method | 6 |
| | A. Equipment | 6 |
| | B. Reagents | 6 |
| | C. Standards | 6 |
| | D. Procedure | 7 |
| | E. Calculations | 8 |
| IV. | Results | 8 |
| | A. Optimization of Chromatography | 8 |
| | B. Linearity | 8 |
| | C. Recovery | 9 |
| V. | Comments | 9 |
| References | | 9 |

## I. INTRODUCTION

Acetaminophen (N-acetyl-p-aminophenol, APAP) is an accepted aspirin substitute which provides analgesic relief through the elevation of the pain threshold. It also reduces fever by acting on the hypothalmic heat-regulating center. In cases where toxic levels are suspected, therapeutic monitoring of this drug is important due to the necrotic effects of overdose concentrations on the hepatic system. This determination can be accomplished using several techniques, including gas chromatography, thin-layer chromatography, spectrophotometry, and liquid chromatography/ultraviolet absorbance detection. The plasma concentrations encountered in APAP overdose are enormous and detection limits are rarely an issue. On the other hand, for basic metabolism studies (including pharmacokinetics) there is a need for better methods. Because of its sensitivity and selectivity, liquid chromatography/electrochemistry (LCEC) provides a method of choice for accurate low level APAP detection (Figure 1). The small volume of serum necessary and the limited amount of sample preparation makes this a fast and simple analysis technique.

## II. PRINCIPLE

Extraction of acetaminophen and the internal standard, propionyl-p-aminophenol, from 100 µℓ of plasma is achieved using ethyl acetate. An aliquot of the organic phase is removed after reciprocal shaking and centrifugation. This sample is evaporated to dryness, reconstituted in mobile phase, filtered, and injected onto a reverse phase column. Sodium perchlorate/sodium citrate buffer-methanol is used as the mobile phase to elute the compounds. Detection is accomplished at +800 mV. The compounds are quantified using peak heights.

## III. MATERIALS AND METHOD

### A. Equipment

A liquid chromatography (LC) system should be equivalent to a Bioanalytical Systems LC-154T (Bioanalytical Systems, Inc., W. Lafayette, Ind.). This system is equipped with an LC-4A/17 electrochemical detector, LC-22/23A temperature controller, and a reverse phase $C_{18}$ Biophase® ODS 5 µm column. The electrochemical detector should employ a TL-5 glassy carbon working electrode.

The nitrogen evaporator used was purchased from Organomation Associates, Northborough, Mass. The centrifuge microfilters, membranes, and microfiltration centrifuge are from Bioanalytical Systems.

### B. Reagents

**Ethyl acetate** — Analytical reagent grade (ACS or equivalent, Fisher Scientific, Fair Lawn, N.J.)

**NaCl** — Mallinckrodt® (Mallinckrodt, Inc., St. Louis)

**Mobile phase** — Add 28.1 g sodium perchlorate (Fisher Scientific), 1.50 g sodium citrate (Fisher Scientific), and 130 mℓ methanol (Mallinckrodt) to a 1 ℓ volumetric flask. Dilute to the mark with distilled water. Adjust pH to 5.0. Filter through a 0.2 µm membrane filter (Rainin Instruments, Woburn, Mass.) and degas before use.

**1.0 M Phosphate buffer** — Dissolve 8.64 g $Na_2HPO_4$ and 2.36 g $KH_2PO_4$ in 100 mℓ of distilled water. Adjust pH to 7.4 using sodium hydroxide.

### C. Standards

Acetaminophen can be purchased from Sigma Chemical Co. (St. Louis) and propionyl-p-aminophenol from Pierce Chemical Co. (Rockford, Ill.).

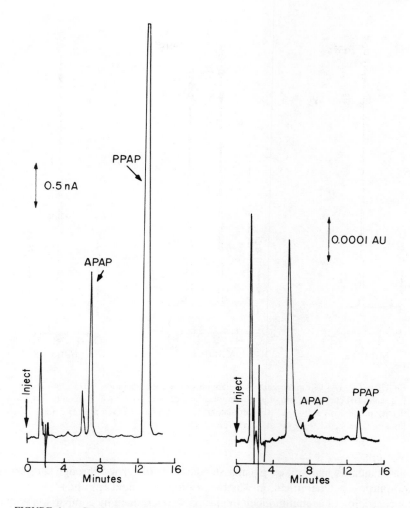

FIGURE 1. Comparison of EC (A) and UV (B) detection (254 nm) of 300 ng/mℓ acetaminophen. The same sample was used for each separation; chromatographic conditions and sample preparation were as described in text.

An acetaminophen standard is prepared by dissolving 25 mg of the compound in 500 mℓ distilled water. This solution is diluted to the appropriate concentration for working standards. All of these stock solutions are stable at 4°C.

Propionyl-*p*-aminophenol standard: dissolve 10 mg of this internal standard in a 100 mℓ volumetric flask with approximately 3 mℓ methanol. Dilute to the mark using 1.0 $M$ phosphate buffer. Dilute this stock solution 50-fold with distilled water and saturate with NaCl for working solution. Store at 4°C.

## D. Procedure

In a 3 mℓ screw-top test tube, add 50 mg NaCl, 100 μℓ working phosphate buffer containing internal standard, 100 μℓ plasma, and 2.5 mℓ ethyl acetate. Cap and shake on a reciprocal shaker for 10 min. Centrifuge 10 min. Transfer 2 mℓ of the organic layer to a 3 mℓ culture tube and evaporate to dryness under nitrogen at 40°C. Reconstitute in 500 mℓ mobile phase. Filter in a centrifugal microfilter through an RC-58 membrane and inject 20 μℓ of this solution.

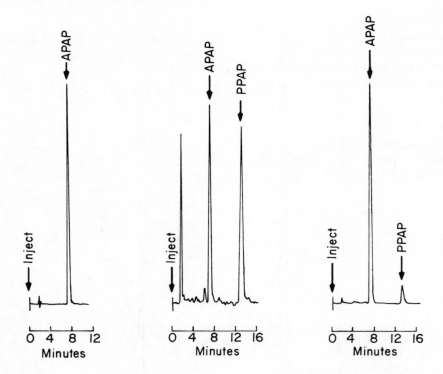

FIGURE 2. (A) Chromatogram of acetaminophen standard solution, 1.0 µg/mℓ at 20 nA full scale; (B) chromatogram of acetaminophen in a plasma extract which initially contained 0.1 µg/mℓ at 2 nA full scale; (C) chromatogram of acetaminophen in plasma extract which initially contained 1.0 µg/mℓ at 20 nA full scale.

### E. Calculations

Aqueous standards and spiked plasma standards were carried through the procedure for calibration purposes. Quantification was based on peak height (pk. ht.)

The concentration of acetaminophen in plasma is determined by comparison with a sample of known concentration which has been carried through the same extraction procedure (Figure 2). Using the following equation, the unknown concentration can be calculated:

$$[APAP]_{unknown} = [APAP]_{known} \times \frac{(Pk.\ ht.\ APAP/Pk.\ ht.\ PPAP)_{unknown}}{(Pk.\ ht.\ APAP/Pk.\ ht.\ PPAP)_{known}}$$

## IV. RESULTS

### A. Optimization of Chromatography

The compounds are eluted with the mobile phase at a flow rate of 1.4 mℓ/min. The column temperature should be maintained at 30°C. Detector potential should be set at +0.80V vs. Ag/AgCl. This relatively low oxidation potential yields improved results because of lower background and noise and fewer chemical interferences. The retention time for acetaminophen under the outlined conditions is 7.2 min and that of the internal standard is 13.2 min. These retention times can be easily varied by adjusting the amount of organic solvent in the mobile phase.

### B. Linearity

The assay is linear for both acetaminophen standards and spiked plasma samples from 2 to 200 ng of injected acetaminophen.

## C. Recovery

The extraction efficiency for this assay was found to be 87.2 ± 3.8% (n = 10) over the established linear concentration range. The minimum detectable concentration was 0.005 µg/mℓ in plasma which yielded a signal-to-noise ratio of 4.5, corresponding to 100 pg of injected acetaminophen.

## V. COMMENTS

Samples carried through the extraction procedure may be refrigerated overnight for next-day analysis without decomposition. The accepted therapeutic range for acetaminophen is 1 to 10 µg/mℓ in plasma with a serum half-life of 2 to 10 hr.

## REFERENCES

1. **Miner, D. J. and Kissinger, P. T.**, Trace determination of acetaminophen in serum, *J. Pharm. Sci.*, 68, 96, 1979.
2. **Pachla, L. A. and Ng, K. T.**, A liquid chromatographic assay for acetaminophen in human plasma utilizing amperometric detection, *J. Chromatogr.*, submitted.

Chapter 3

# DETERMINATION OF ϵ-AMINOCAPROIC ACID

## Gary J. Schmidt

## TABLE OF CONTENTS

I. Introduction ................................................................... 12

II. Principle ...................................................................... 12

III. Materials and Methods ........................................................ 12
    A. Equipment ............................................................... 12
    B. Reagents ................................................................. 12
    C. Standards ................................................................ 12
    D. Procedure ................................................................ 12
    E. Calculation and Calibration ............................................. 13

IV. Results ....................................................................... 14
    A. Linearity ................................................................ 14
    B. Recovery ................................................................. 14
    C. Detection ................................................................ 14
    D. Precision ................................................................ 14
    E. Accuracy ................................................................. 14
    F. Interferences ............................................................ 14
    G. Patient Serum ........................................................... 14

V. Comment ...................................................................... 14

Reference ......................................................................... 15

## I. INTRODUCTION

ε-Aminocaproic acid is an important antifibrinolytic agent. A simple liquid chromatography procedure is described for determining this drug in 10 µℓ of serum. The procedure requires 15 min and is useful for determining the drug at therapeutic concentrations from 100 to 400 mg/ℓ.

## II. PRINCIPLE

A 10-µℓ volume of serum is deproteinized with ethanol. The supernatant is evaporated to dryness and derivatized with dansyl chloride reagent for 15 min. An aliquot of the derivatized sample is analyzed on a $C_8$ reversed-phase column and the column effluent monitored using a fluorescence detector.

## III. MATERIALS AND METHODS

### A. Equipment

A liquid chromatography (LC) system equivalent to the following is used: a Model Series 2/2 liquid chromatograph (Perkin-Elmer Corp., Norwalk, Conn.) equipped with a Rheodyne® 7105 valve (Rheodyne, Cotati, Calif.), a fluorescence detector (Perkin-Elmer Model 650 LC or equivalent), a column oven (Perkin-Elmer Model LC-100), and a strip-chart recorder (Perkin-Elmer Model 56). The column is a $C_8$ (octysilica) reversed-phase 25 cm × 4.6 mm i.d. (Perkin-Elmer) or equivalent. Special glassware includes 5-mℓ conical glass centrifuge tubes, 2-mℓ glass sample vials with open top screw-caps with polytetrafluoroethylene lined septa (Precision Sampling, Baton Rouge, La.), and 10, 20, and 50-µℓ disposable glass pipets. A bench-top centrifuge, a block heater, an evaporation manifold with a source of dry air, and a vortex-mixer are also used.

### B. Reagents

Methanol and acetone (LC Grade), distilled in glass (Burdick and Jackson Laboratories Inc., Muskegon, Mich.). Dansyl chloride (100 g/ℓ in acetone) (Pierce Chemical Co., Rockford, Ill.). Ethanol (absolute) (Publicker Industries Inc., Philadelphia, Pa.). Dansyl chloride working solution: dilute 0.1 mℓ dansyl chloride stock solution to a volume of 8 mℓ with acetone. Bicarbonate buffer: 0.1 mol/ℓ sodium bicarbonate. Filter through a 0.45 µm Millipore® filter (Millipore Corp., Bedford, Mass.). Mobile phase: prepare by mixing methanol/water/$H_3PO_4$ in the proportions 40/60/0.1 by volume and degas.

### C. Standards

ε-Aminocaproic acid and norleucine can be obtained from Sigma Chemical Co., St. Louis. Individual standard solutions of ε-aminocaproic acid and norleucine are prepared in 0.01 $N$ HCl to a final concentration of 2 g/ℓ. Solutions are stored at 4°C.

**Serum standards** — Serum standards are prepared from a blank serum pool to contain ε-aminocaproic acid at concentrations of 100, 200, and 400 mg/ℓ.

### D. Procedure

To 10 µℓ of serum sample or serum standard in a 5-mℓ glass centrifuge tube, add 40 µℓ ethanol and 2 µℓ internal standard solution. Mix vigorously for 15 sec on a vortex-type mixer and centrifuge for 1 min at 2000 r/min (710 g). Transfer the supernatant to the bottom of a 2-mℓ sample vial and evaporate to dryness at 70°C with gentle current of air. Add 10 µℓ of bicarbonate buffer and 40 µℓ of dansyl chloride working solution. Cap the vial tightly and mix vigorously for 15 sec. Heat the vial at 70°C for 15 min in a block heater. Remove

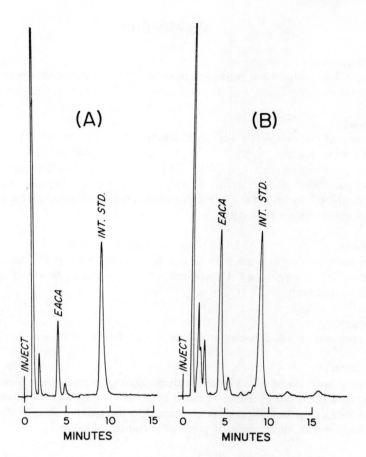

FIGURE 1. (A) Chromatogram of derivatized standards of ε-aminocaproic acid (EACA) and norleucine (Int. Std.); (B) serum standard with ε-aminocaproic acid at concentration of 200 mg/ℓ. (From Adams, R. F., Schmidt, G. J., and Vandemark, F. L., *Clin. Chem.*, 23, 1226, 1977. With permission.)

the vial, cool, and add 50 μℓ acetone. Inject a 1-μℓ aliquot into the chromatograph. The mobile phase flow rate is 1.5 mℓ/min and the column temperature is 60°C. The fluorescence detector is set at excitation and emission wavelengths of 345 and 545 nm, respectively, using 10-nm bandwidths. Chromatograms illustrating the separation of ε-aminocaproic acid and norleucine are shown in Figure 1. Chromatogram A is a standard of 10 ng ε-aminocaproic acid and 40 ng norleucine. Chromatogram B is from the analysis of a serum standard containing 200 mg/ℓ ε-aminocaproic acid.

### E. Calculation and Calibration

Serum standards are processed by the procedure and a standard curve based on peak area ratios of drug to internal standard vs. concentration is prepared. The patients' sera are then analyzed, the peak areas ratios calculated, and the ε-aminocaproic acid concentration determined from the working curve.

## IV. RESULTS

### A. Linearity
A linear relationship of the peak area ratios exist over the tested concentration range of 50 to 1000 mg/$\ell$.

### B. Recovery
Recoveries of $\epsilon$-aminocaproic acid from serum over the concentration range of 50 to 1000 mg/$\ell$ were 96 to 102%.

### C. Detection
Using the 10-$\mu\ell$ serum volume as recommended, $\epsilon$-aminocaproic acid may be determined to a lower concentration of 4 mg/$\ell$.

### D. Precision
Within-run precision at $\epsilon$-aminocaproic acid concentrations of 100 and 250 mg/$\ell$ were 4.2 and 3.1 CV%, respectively. Day-to-day precision at these concentrations were 6.1 and 4.9 CV%, respectively.

### E. Accuracy
Correlation with a gas chromatography procedure for this drug was r = 0.986.

### F. Interferences
No natural amino acid tested eluted with norleucine. Methionine elutes close to $\epsilon$-aminocaproic acid and could interfere if column efficiency is impaired. However, normal serum concentrations of methionine are low compared to therapeutic concentrations of $\epsilon$-aminocaproic acid, and an interference, if present, would be tolerable.

### G. Patient Serum
Figure 2 indicates the performance obtained when analyzing patient serum samples. Chromatogram A shows a serum sample having an $\epsilon$-aminocaproic acid concentration of 98 mg/$\ell$ and chromatogram B has an $\epsilon$-aminocaproic acid concentration of 181 mg/$\ell$.

## V. COMMENT

The therapeutic concentration range of $\epsilon$-aminocaproic appears to be in the range of 100 to 400 mg/$\ell$.

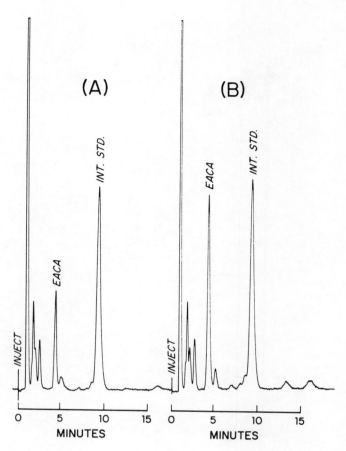

FIGURE 2. Patient sera processed according to the procedure. (A) 98 mg/ℓ; (B) 181 mg/ℓ. (From Adams, R. F., Schmidt, G. J., and Vandemark, F. L., *Clin. Chem.*, 23, 1226, 1977. With permission.)

# REFERENCE

1. **Adams, R. F., Schmidt, G. J., and Vandemark, F. L.,** Determination of ε-aminocaproic acid in serum by reversed-phase chromatography with fluorescence detection, *Clin. Chem.*, 23, 1226, 1977.

Chapter 4

# ANALYSIS OF AMINOGLYCOSIDE ANTIBIOTICS BY PRECOLUMN FLUORESCENCE DERIVATIZATION

**Wayne L. Settle and Martha R. Harkey**

## TABLE OF CONTENTS

I. Introduction ................................................................. 18

II. Principle ..................................................................... 18

III. Materials and Methods .................................................. 18
    A. Equipment ............................................................. 18
    B. Reagents ................................................................ 18
    C. Standards .............................................................. 19
    D. Procedure .............................................................. 19
    E. Calculation ............................................................ 19

IV. Results ....................................................................... 21
    A. Linearity ............................................................... 21
    B. Recovery and Stability ............................................ 21
    C. Interferences .......................................................... 21
    D. Precision ............................................................... 21
    E. Accuracy ............................................................... 21
    F. Temperature .......................................................... 21
    G. Correlation with Immunoassay ................................ 21

V. Discussion .................................................................. 21

VI. Comments .................................................................. 22

References ........................................................................ 24

## I. INTRODUCTION

Aminoglycoside antibiotics used in the treatment of Gram-negative bacterial infections have the potential for nephrotoxicity and ototoxicity if not carefully monitored.[1-3] Since these drugs are eliminated unchanged by the kidneys,[4] it is especially important to follow the serum levels in patients with reduced renal function[5,6] or in burn patients who generally require higher doses to maintain a therapeutic level.[7]

There are several published techniques using liquid chromatography of fluorescent derivatives to assay aminoglycoside antibiotics.[8-12] Since these drugs have very little inherent absorbance or fluorescence, they must be derivatized either before or after chromatography. Precolumn derivatization is cost effective for the analysis of aminoglycoside antibiotics since the derivatization can be combined with the sample clean up and eliminates the need for a postcolumn reagent pump and reaction coil. Precolumn derivatization has the disadvantage, however, that multiple derivatives with varying stabilities may be formed.[13]

Of the aminoglycoside antibiotics in current use, most therapeutic drug monitoring programs have centered on gentamicin and tobramycin and a variety of immunoassay techniques have been developed for the analysis of these two drugs. Although the chromatographic procedure described in this chapter was developed for gentamicin and tobramycin, it may be applied to any aminoglycoside antibiotic including amikacin, kanamycin, and the newer drugs, netilmicin and sisomicin.

## II. PRINCIPLE

Serum or plasma proteins are removed from the sample by adsorbing the aminoglycosides onto silica gel and washing off the serum proteins with water. The derivatizing reagent is added and the drug is extracted from the silica gel with ethanol. A stable derivative is insured by heating at 50°C, then cooling to 0°C in ice. An aliquot of the sample is injected onto a reversed phase column and the drugs are eluted with acetonitrile-acetate buffer mobile phase. The drugs are detected by the fluorescence of their major derivative and quantified from their relative peak heights or areas.

## III. MATERIALS AND METHODS

### A. Equipment

A liquid chromatographic (LC) system equivalent to the following should be used: a Series 1 pump (Perkin-Elmer, Norwalk, Conn.) equipped with a Rheodyne® 7105 sample injector and model 7302 column inlet filter (Rheodyne, Cotati, Calif.), a model 204-A fluorescence spectrophotometer (Perkin-Elmer), a column jacket and water bath capable of holding the column temperature at 35 ± 1°C, and an Ultrasphere® octyl column, 15 cm × 4.6 mm i.d. (Altex Scientific, Berkeley, Calif.). A strip-chart recorder and/or a Sigma® 10 data system (Perkin-Elmer), a model 5432 Eppendorf® shaker and model 5412 Eppendorf® centrifuge (Brinkman Instruments, Inc., Westbury, N.Y.) and 1.5 mℓ polypropylene tubes are also used.

### B. Reagents

Acetonitrile is HPLC grade (J. T. Baker, Phillipsburg, N.J.) and ethanol is spectrophotometric grade (J. T. Baker). All other chemicals are reagent grade. Water is deionized and glass-distilled daily. Silica gel 60-200 mesh (J. T. Baker) is washed three times each with water, ethanol, and methanol, respectively. It is then air dried and stored in a closed glass container.

The o-phthalaldehyde reagent is identical to that described by Bäck et al.[11] and Maitra et al.[12] — 1 g of boric acid is dissolved in 38 m$\ell$ of water and adjusted to pH 10.4 with 45% potassium hydroxide. There are 2 m$\ell$ of methanol used to dissolve 200 mg of o-phthalaldehyde (Pickering Laboratories, Mountain View, Calif.), then added to the boric acid solution along with 0.4 m$\ell$ 2-mercaptoethanol (Sigma Chemical Co., St. Louis). The reagent is stored under nitrogen at 4°C and prepared weekly.

Sodium acetate buffer, pH 3.5, is prepared by diluting 4.0 g anhydrous sodium acetate and 40 m$\ell$ glacial acetic acid to 1 $\ell$ with water. The optimum mobile phase must be adjusted for each column and modified slightly as the column ages (see Section V). The chromatograms shown here are with mobile phases of either 550 m$\ell$ acetonitrile (for gentamicin) or 520 m$\ell$ acetonitrile (for tobramycin or amikacin) and 260 m$\ell$ 750 mM sodium acetate, pH 3.5, diluted to 1 $\ell$ with distilled water, then filtered through a 0.45 μm nylon filter (Rainin Instrument Co., Woburn, Mass.).

## C. Standards

Gentamicin sulfate, sisomicin sulfate, netilmicin sulfate, and purified gentamicin components $C_1$, $C_{1a}$, and $C_2$ may be obtained from Schering Drug Corporation, Bloomfield, N.J., tobramycin from Eli Lilly and Company, Indianapolis, Ind., and amikacin and kanamycin from Bristol Laboratories, Syracuse, N.Y.

Stock solutions prepared at a concentration of 1 mg/m$\ell$ in water are stable at 4°C for at least 6 months. These solutions are diluted with drug-free serum or water to obtain drug concentrations of 1, 5, 10, and 20 mg/$\ell$ for the serum or reference standards and 0.5 m$\ell$ aliquots of these solutions are stored at 20°C. These are stable for 6 months.

The internal standard, sisomicin, is prepared daily by diluting the stock solution with water to a concentration of 5.0 mg/$\ell$.

## D. Procedure

To a 1.5 m$\ell$ polypropylene tube, add 150 μ$\ell$ of the washed silica gel (may be measured with a scoop prepared from a 1.0 m$\ell$ tuberculin syringe), 1 m$\ell$ water, 100 μ$\ell$ internal standard, and 200 μ$\ell$ serum sample, serum standard, or control. For tobramycin and amikacin, smaller sample volumes (100 μ$\ell$) may be used and the internal standard omitted.* Mix 1 min in an Eppendorf® shaker, then centrifuge 2 min at 10,000 × g in an Eppendorf® centrifuge. After discarding the supernatant, add 200 μ$\ell$ o-phthalaldehyde reagent and 500 μ$\ell$ absolute ethanol to the silica gel. Mix 1 min and centrifuge 2 min. Without disturbing the silica gel pellet, transfer 500 μ$\ell$ of the supernatant to an opaque 1.5 m$\ell$ polypropylene tube to protect from light. After setting for 10 min at 50°C, store the sample on ice, and inject 40 μ$\ell$ onto the column maintained at 35°C. The sample is eluted at a flow rate of 3.0 m$\ell$/min and monitored at an excitation wavelength of 335 nm and emission wavelength of 430 nm. Photomultiplier gain is set at 4 with sensitivity range at 3 and excitation and emission slits at 10 nm.

Figure 1 shows chromatograms of a gentamicin serum standard and patient sample on gentamicin therapy. The chromatograms for tobramycin standard and patient are shown in Figure 2.

## E. Calculation

A standard curve is constructed daily from duplicate injections of a 1, 5, and 10 mg/$\ell$ sample of gentamicin or tobramycin in serum. For amikacin or kanamycin, standards ranging to 35 mg/$\ell$ should be used.

---

* The internal standard, sisomicin, has a retention time of over 18 min in the tobramycin mobile phase. This lengthens the assay unnecessarily with no improvement in precision.

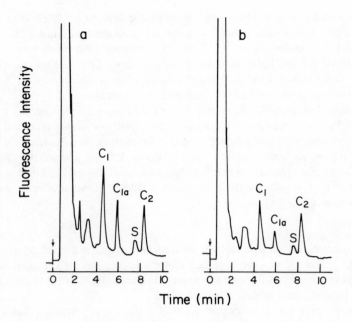

FIGURE 1. Chromatogram of serum extracts of gentamicin. (a) Serum standard at 10 mg/$\ell$; (b) patient sample at 6.1 mg/$\ell$.

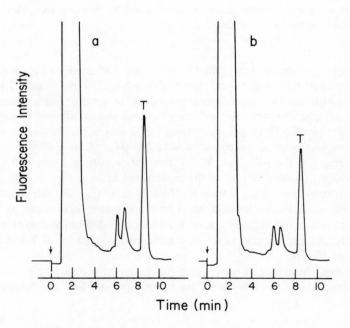

FIGURE 2. Chromatogram of serum extracts of tobramycin. (a) Serum standard at 5 mg/$\ell$; (b) patient sample at 4.0 mg/$\ell$.

The aminoglycoside concentration in patient samples is calculated by comparison of the peak heights or areas to those obtained from the serum standards.

## IV. RESULTS

### A. Linearity
Using the optimal amount of derivatizing reagent (200 $\mu\ell$), a maximum fluorescence is produced with linearity from 1 to 20 mg/$\ell$ gentamicin and 0.3 to 20 mg/$\ell$ tobramycin.

### B. Recovery and Stability
The absolute recovery for the various aminoglycosides in serum is as follows: gentamicin components $C_1$, 1.00, $C_{1a}$, 0.99, and $C_2$, 1.04; sisomicin, 1.05; tobramycin, 0.80; amikacin, 0.68; and netilmicin 0.92. No difference in the recovery from serum or water is observed. The stability of derivatives for all the aminoglycosides are similar with a decay of 2 to 4%/hr at 4°C.

### C. Interferences
No interferences have been observed with this assay. Drugs which fluoresce, such as quinidine and propranol, have much shorter retention times than the aminoglycosides. By using the two mobile phases (one for gentamicin, netilmicin, and sisomicin; the other for amikacin, tobramycin, and kanamycin), all aminoglycosides may be resolved.

If the silica gel is not washed thoroughly or stored in plastic containers, extraneous peaks may be observed which could interfere with the quantitation of tobramycin.

### D. Precision
Samples at 1 mg/$\ell$, extracted and derivatized separately on the same day, gave coefficients of variation of 8% for gentamicin and 3% for tobramycin. At 10 mg/$\ell$, the coefficients of variation were 5 and 2% for gentamicin and tobramycin, respectively. Day to day variations were 8% for gentamicin and 5% for tobramycin. The variability of gentamicin was greater because three component peaks were included in quantitation.

### E. Accuracy
Linear regression analysis of gentamicin concentrations for standard samples in human serum over 6 months was represented by the equation $y = 1.06x + 0.01$, with a correlation coefficient, $r = 0.975$. Over the same period linear regression analysis for tobramycin standard samples was represented by $y = 1.00x + 0.25$ with $r = 0.951$.

### F. Temperature
The retention times and fluorescence of the derivatized aminoglycosides were strongly temperature dependent. There was a gradual decrease in fluorescence as column temperature was increased to 45°C, where approximately 50% loss of fluorescence was observed. By maintaining the column at 35°C, the retention times were stabilized and the magnitude of the fluorescence provided the required sensitivity. If the laboratory temperature is constant, the assay may be performed at room temperature.

### G. Correlation with Immunoassay
This liquid chromatographic procedure has been compared to fluorescence immunoassay (Abbott $TD_x$) using 32 serum samples from patients on tobramycin therapy. A linear regression analysis of the data gives the equation $y = 0.9254x + 0.4612$ with a correlation coefficient, $r = 0.9103$.

## V. DISCUSSION

The retention times of the various aminoglycosides are affected by the percentage of acetonitrile and the ionic strength and pH of the buffer. The shapes of these curves are

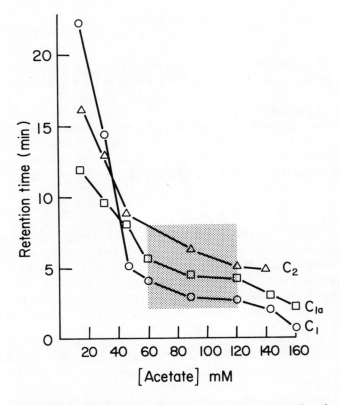

FIGURE 3. Retention time dependence on acetate concentration of derivatized samples of purified gentamicin components. ○ = $C_1$, □ = $C_{1a}$, △ = $C_2$.

similar for different columns; however, the scale of the abscissa varies from column to column. The ionic strength of the mobile phase should be adjusted so that one is operating within the shaded area shown in Figure 3. After a column has seen approximately 1000 injections, the $C_1$ and $C_{1a}$ components of gentamicin move closer together. The resolution of these peaks can be improved by increasing the sodium acetate concentration in the mobile phase.

As shown in Figure 4, the optimum pH is in the range of 3 to 6 pH units. The fluorescence of the aminoglycoside derivatives decrease rapidly below pH 3.5. Some aminoglycosides, e.g., netilmicin, exhibit severe tailing which can be minimized by lowering the pH of the mobile phase.

The excitation and emission spectra for both gentamicin and the mobile phase are shown in Figure 5. There is a large excitation peak which begins near the peak of the aminoglycosides. The large peak, however, produces no emission at 430 nm, thus it does not interfere with the assay using 335 nm excitation wavelength and 430 nm emission wavelength.

## VI. COMMENTS

For proper dosage adjustment, both peak and trough levels of aminoglycoside antibiotics are monitored. The peak level is obtained 15 to 30 min after the end of an i.v. infusion and the trough is drawn just before the next dose. In addition, accurate notation of both the sample time and time of last dose are required for pharmacokinetic interpretation.

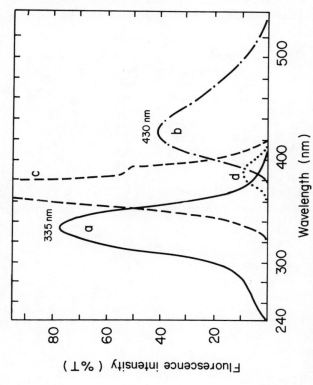

FIGURE 5. Excitation and emission spectra of gentamicin. (a) Excitation spectrum of derivatized gentamicin; (b) emission spectrum of derivatized gentamicin; (c) excitation spectrum of mobile phase; (d) excitation spectrum of water blank. The mobile phase had no emission fluorescence which would interfere with the emission spectrum of gentamicin.

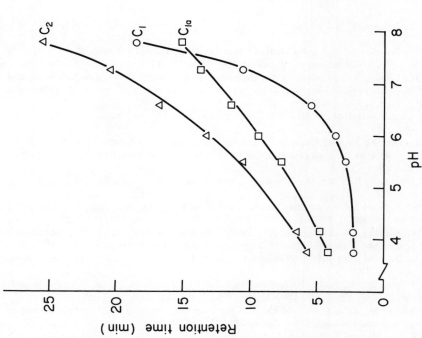

FIGURE 4. Retention time dependence on pH of derivatized samples of purified gentamicin components. ○ = $C_1$, □ = $C_{1a}$, △ = $C_2$.

Although the desired therapeutic level varies with the type and severity of infection, the following levels are generally associated with therapeutic response and increased risk of toxicity:

|  |  | Therapeutic level (mg/ℓ) | Toxic level (mg/ℓ) |
|---|---|---|---|
| Gentamicin or tobramycin | Peak | 4—10 | >12 |
|  | Trough | 0.5—1.5 | >2 |
| Amikacin | Peak | 20—30 | >35 |
|  | Trough | 1—4 | >10 |

The therapeutic and toxic levels for netilmicin and sisomicin will be similar to those for gentamicin and tobramycin, whereas the therapeutic levels for kanamycin are the same as amikacin.

## REFERENCES

1. **Barza, M. and Lauerman, M.,** Why monitor serum levels of gentamicin?, *Clin. Pharmacokinet.*, 3, 202, 1978.
2. **Smith, C. R., Baughman, K. L., Edwards, C. Q., Rogers, J. F., and Lietman, P. S.,** Controlled comparison of amikacin and gentamicin, *N. Engl. J. Med.*, 296, 349, 1977.
3. **Smith, C. R., Lipsky, J. J., Laskin, O. L., Hellmann, D. B., Mellits, E. D., Longstreth, J., and Lietman, P. S.,** Double-blind comparison of the nephrotoxicity and auditory toxicity of gentamicin and tobramycin, *N. Engl. J. Med.*, 302, 1106, 1980.
4. **Pechere, J. C. and Dugal, R.,** Clinical pharmacokinetics of aminoglycoside antibiotics, *Clin. Pharmacokinet.*, 4, 170, 1979.
5. **Kaye, D., Levison, M. E., and Labovitz, E. D.,** The unpredictability of serum concentrations of gentamicin: pharmacokinetics of gentamicin in patients with normal and abnormal renal function, *J. Infect. Dis.*, 130, 150, 1974.
6. **Barza, M., Brown, R. B., Shen, D., Gibaldi, M., and Weinstein, L.,** Predictability of blood levels of gentamicin in man, *J. Infect. Dis.*, 132, 165, 1975.
7. **Zaske, D. E., Sawchuk, R. J., Gerding, D. N., and Strate, R. G.,** Increased dosage requirements of gentamicin in burn patients, *J. Trauma*, 16, 824, 1976.
8. **Peng, G. W., Gadalla, M. A. F., Peng, A., Smith, V., and Chiou, W. L.,** High-pressure liquid-chromatographic method for determination of gentamicin in plasma, *Clin. Chem.*, 23, 1838, 1977.
9. **Anhalt, J. P.,** Assay of gentamicin in serum by high-pressure liquid chromatography, *Antimicrob. Agents Chemother.*, 11, 651, 1977.
10. **Anhalt, J. P. and Brown, S. D.,** High-performance liquid-chromatographic assay of aminoglycoside antibiotics in serum, *Clin. Chem.*, 24, 1940, 1978.
11. **Bäck, S., Nilsson-Ehle, I., and Nilsson-Ehle, P.,** Chemical assay, involving liquid chromatography, for aminoglycoside antibiotics in serum, *Clin. Chem.*, 25, 1222, 1979.
12. **Maitra, S. K., Yoshikawa, T. T., Hansen, J. L., Nilsson-Ehle, I., Palin, W. J., Schotz, M. C., and Guze, L. B.,** Serum gentamicin assay by high-performance liquid chromatography, *Clin. Chem.*, 23, 2275, 1977.
13. **Cooper, D. J., Weinstein, J., and Waitz, J. A.,** Gentamicin antibiotics. 4. Some condensation products of gentamicin $C_2$ with aromatic and aliphatic aldehydes, *J. Med. Chem.*, 14, 1118, 1971.

Chapter 5

# SERUM AMIKACIN WITH SPECTROPHOTOMETRIC DETECTION

## Pokar M. Kabra

## TABLE OF CONTENTS

I. Principle .................................................................... 26

II. Materials and Methods ..................................................... 26
    A. Equipment ............................................................. 26
    B. Reagents .............................................................. 26
    C. Standards ............................................................. 26
    D. Procedure ............................................................. 26
    E. Calculations .......................................................... 28

III. Results .................................................................... 28
    A. Linearity ............................................................. 28
    B. Recovery ............................................................. 28
    C. Precision ............................................................. 28
    D. Interference .......................................................... 28
    E. Accuracy ............................................................. 28

IV. Comments .................................................................. 28

Reference ....................................................................... 28

## I. PRINCIPLE

Amikacin and kanamycin (internal standard) are converted into trinitrophenyl derivatives by reacting them with a water soluble derivatizing agent (2,4,6-trinitrobenzene sulfonic acid, TNBS) at 70°C for 30 min. The trinitrophenyl derivatives are extracted from the crude mixture with a reversed-phase Bond-Elut® $C_{18}$ column. An aliquot of the extracted sample is injected onto a reversed-phase octyl column and eluted with acetonitrile-phosphate buffer mobile phase. The eluted derivatives are monitored at 340 nm and quantitated from their peak height or peak areas.

## II. MATERIALS AND METHODS

### A. Equipment

Liquid chromatography (LC) systems equivalent to the following are recommended: a Model Series 2 or Series 3 liquid-chromatograph equipped with a 7105 injection valve, a Model 75 or 65T variable wavelength detector, a Model LC 100 temperature controlled oven (all from Perkin-Elmer Corp., Norwalk, Conn.), a reversed-phase 25 cm × 4.6 mm Ultrasphere® octyl 5 μm column (Altex Scientific, Berkeley, Calif.) mounted in the oven, and a Sigma® 10 data system (Perkin-Elmer) are used. A Model 5412 Eppendorf® centrifuge with 1.5 mℓ polypropylene tubes (Brinkman Instruments, Inc., Westbury, N.Y.) are used. Vac-Elut® vacuum chamber and Bond-Elut® $C_{18}$ extraction columns (Analytichem International, Inc., Harbor City, Calif.) are used to extract the amikacin derivative.

### B. Reagents

All chemicals used are of reagent grade. Acetonitrile and methanol distilled in glass are from Burdick and Jackson Laboratories, Inc., Muskegon, Mich. Phosphate buffer, 20 mmol/ℓ is prepared by dissolving 1.34 g of $KH_2PO_4$ in 500 mℓ of water. This solution is titrated to pH 3.0 with phosphoric acid. Mobile phase: This is a solution of 520 mℓ acetonitrile in 480 mℓ 20 mmol/ℓ phosphate buffer, pH 3.0. Tris buffer, 2 mol/ℓ, pH 10.3 is prepared by dissolving 24.2 g of Trizma® base (Sigma Chemical Co., St. Louis) in 100 mℓ of water. This solution is stable at 4°C for at least 1 year. Wash buffer, 10 mmol/ℓ, pH 8.6, is prepared by dissolving 1.74 g of $K_2HPO_4$ in 1 ℓ of water. This solution is titrated to pH 8.6 with phosphoric acid. The buffer is stable at ambient temperature for at least 1 year.

### C. Standards

Amikacin and kanamycin sulfate can be obtained from Bristol Laboratories, Syracuse, N.Y. The amikacin stock standard is prepared by dissolving 25 mg in 100 mℓ of water. The solution is stable at 4°C for at least 6 months. The working serum standards (5, 10, and 25 μg/mℓ) are prepared by adding 200, 400, and 1000 μℓ of stock standard to 9.8, 9.6, and 9.0 mℓ of drug-free serum. The serum standards are stable for at least 1 month at 4°C. The stock internal standard is prepared by dissolving 25 mg of kanamycin in 100 mℓ of acetonitrile. The solution is stable at 4°C for at least 6 months. The working internal standard is prepared by diluting the stock solution tenfold with acetonitrile. This solution is stable at 4°C for at least 1 month.

### D. Procedure

To 50 μℓ of serum sample, serum standard, or control in 1.5 mℓ polypropylene tube, add 25 μℓ of 2 mol/ℓ tris buffer and 100 μℓ of working internal standard (kanamycin, 25 μg/mℓ). Vortex-mix and centrifuge for 1 min in Eppendorf® centrifuge at 15,000 g. Decant the supernate into a second set of appropriately labeled polypropylene tubes and add 30 μℓ of TNBS solution. Cap, vortex, and place the tubes in a 70°C heating block for 30 min. For each sample, place a Bond-Elut® $C_{18}$ extraction column on top of the Vac-Elut® chamber

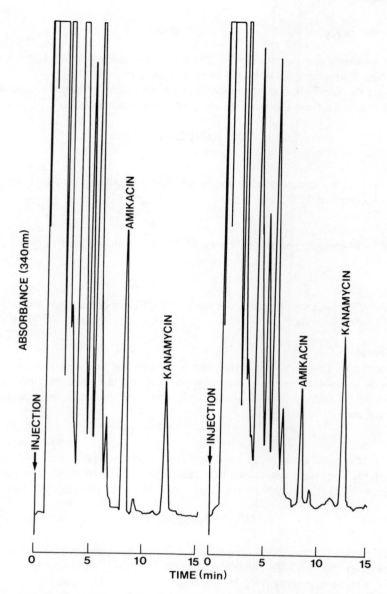

FIGURE 1. Left: chromatogram of a patient's serum containing 26.2 µg/mℓ of amikacin; right: chromatogram of a patient's serum containing 8.6 µg/mℓ of amikacin.

before connecting to vacuum. Pass two column volumes of methanol followed by two column volumes of water through each column. Disconnect vacuum and fill each column with 700 µℓ of working wash solution followed by approximately 200 µℓ of derivatized sample. Connect vacuum to the chamber and pass three volumes of working wash solution through each column.

Disconnect vacuum and place a rack containing labeled 10 × 75 mm glass tubes in the Vac-Elut® chamber. Pipette 300 µℓ of acetonitrile onto each column and connect vacuum. Remove the rack of tubes from the vacuum chamber, shake the tubes to mix the eluate, and inject 50 µℓ of each eluate onto the liquid chromatograph. The column is eluted with the mobile phase at a flow rate of 2.0 mℓ/min. The column is maintained at 50°C, and the column effluent is monitored at 340 nm (Figure 1).

### E. Calculations

1. Serum standards are carried through the procedure and used to prepare a standard curve based on relative peak height or peak area ratios.
2. Calculate the concentration of amikacin in the specimen by direct comparison with the data obtained from the standard curve of the serum standards.

## III. RESULTS

### A. Linearity[1]

Drug concentrations and peak area ratios are linear to 50 mg/$\ell$. The lower limit of sensitivity is approximately 0.5 mg/$\ell$.

### B. Recovery[1]

Analytical recoveries ranged from 94.0 to 98.5% in a concentration range of 2.5 to 50 mg/$\ell$.

### C. Precision[1]

The precision for amikacin equals less than 5% coefficient of variation for both within-day and day-to-day analysis.

### D. Interference[1]

None of the more than 30 drugs tested interfered with the analysis of amikacin. Other aminoglycoside antibiotics such as gentamicin and tobramycin elute after amikacin and do not interfere with the analyses. Grossly hemolyzed, icteric or lipemic samples can be assayed without interference.

### E. Accuracy[1]

The results of our LC assay were compared with a commercially available radioimmunoassay method. The linear regression data comparing the LC method with radioimmunoassay were n = 25, r = 0.999, y-intercept = 0.398 mg/$\ell$, and slope = 1.047.

## IV. COMMENTS

The conditions for derivatization, extraction, and chromatography were carefully investigated. A large excess of TNBS (18,000 to 20,000 $M$ ratio) is necessary to yield a single derivative for amikacin quantitatively in less than 30 min at 70°C. The pH of the reaction mixture is very critical for complete derivatization. The optimum pH range for this reaction is between 9.5 to 10. The derivatives are extracted from the crude mixture with a Bond-Elut® extraction column. These columns selectively retain the nonpolar amikacin derivative, while excess reagent and polar compounds are washed off with the phosphate buffer.

## REFERENCE

1. **Kabra, P. M., Bhatnagar, P. K., Nelson, M. A., and Marton, L. J.,** Liquid chromatographic determination of amikacin in serum with spectrophotometric detection, *J. Chromatogr.*, submitted.

Chapter 6

# SERUM GENTAMICIN WITH SPECTROPHOTOMETRIC DETECTION

**Pokar M. Kabra**

## TABLE OF CONTENTS

I. Principle .................................................................................. 30

II. Materials and Methods ............................................................ 30
    A. Equipment ...................................................................... 30
    B. Reagents ......................................................................... 30
    C. Standards ........................................................................ 30
    D. Procedure ....................................................................... 30
    E. Calculations .................................................................... 31

III. Results ..................................................................................... 31
    A. Linearity ......................................................................... 31
    B. Recovery ........................................................................ 32
    C. Precision ........................................................................ 32
    D. Interference .................................................................... 32
    E. Accuracy ........................................................................ 32

IV. Comments ............................................................................... 32

Reference ............................................................................................ 32

## I. PRINCIPLE

Gentamicin and sisomicin (internal standard) are converted into trinitrophenyl derivatives by reacting them with a water soluble derivatizing agent (2,4,6-trinitrobenzene sulfonic acid, TNBS) at 70°C for 30 min. The gentamicin and internal standard derivatives are extracted from the crude mixture with chloroform, the solvent is evaporated to dryness and the residue reconstituted in 100 µℓ of acetonitrile. An aliquot is injected onto a reversed-phase octyl column and eluted with acetonitrile phosphate buffer mobile phase. The eluted derivatives are monitored at 340 nm and quantitated from their peak height or peak areas.

## II. MATERIALS AND METHODS

### A. Equipment

Liquid chromatography (LC) systems equivalent to the following are recommended: a Model Series 2 or Series 3 liquid chromatograph equipped with a 7105 injection valve, a Model 75 or 65T variable wavelength detector, a Model LC 100 temperature controlled oven (all from Perkin-Elmer Corp., Norwalk, Conn.), a reversed-phase 15 cm × 4.6 mm Ultrasphere® octyl 5 µm column (Altex Scientific, Berkeley, Calif.) mounted in the oven, and a Sigma® 10 data system (Perkin-Elmer) are used. A Model 5412 Eppendorf® centrifuge with 1.5 mℓ polypropylene tubes (Brinkman Instruments, Inc., Westbury, N.Y.), and an Evapo-Mix® (Buchler Instruments, Fort Lee, N.J.) are utilized.

### B. Reagents

All chemicals used are of reagent grade. Acetonitrile, methanol, and chloroform, all distilled in glass, are from Burdick and Jackson Laboratories, Inc., Muskegon, Mich. Phosphate buffer, 20 mmol/ℓ is prepared by dissolving 2.70 g of $KH_2PO_4$ in 1 ℓ of water. The solution is titrated to pH 3.0 with phosphoric acid. Mobile phase: this is a solution of 660 mℓ of acetonitrile in 340 mℓ of 20 mmol/ℓ phosphate buffer. Tris buffer, 2 mol/ℓ, pH 10.3, is prepared by dissolving 24.2 g of Trizma® base (Sigma Chemical Co., St. Louis) in 100 mℓ of water. This solution is stable for at least 1 year at 4°C. 2,4,6 Trinitrobenzene-1-sulfonic acid (TNBS, Sigma) 250 g/ℓ is prepared by dissolving 2.5 g of TNBS in 10 mℓ of distilled water. This solution is stable for 1 month at 4°C.

### C. Standards

Gentamicin sulfate can be obtained from Sigma Chemical Co., gentamicin $C_1$, $C_{1a}$, and $C_2$ isomers and sisomicin can be obtained from Schering Corp., Bloomfield, N.J.

The gentamicin stock standard is prepared by dissolving 25 mg equivalent of gentamicin activity in 100 mℓ of water. The stock solution is stable for at least 6 months at 4°C. The working gentamicin serum standards (2.5 and 10 mg/ℓ) are prepared by diluting 100 and 400 µℓ of stock standard with 9.9 and 9.6 mℓ of drug-free serum, respectively. Serum standards are stable for at least 1 month at 4°C.

The stock internal standard is prepared by dissolving 25 mg of sisomicin in 100 mℓ of acetonitrile. This solution is stable at 4°C for at least 6 months. The working internal standard is prepared by diluting 2 mℓ of stock internal standard with 98 mℓ of acetonitrile. The solution is stable at 4°C for 1 month.

### D. Procedure

To 50 µℓ of serum sample, standard, and control in a 1.5 mℓ polypropylene tube, add 20 µℓ of 2 mol/ℓ tris buffer, and 100 µℓ of working internal standard (sisomicin, 5 mg/ℓ). Vortex-mix and centrifuge for 1 min in Eppendorf® centrifuge at 15,000 g. Decant the supernate into a second set of appropriately labeled polypropylene tubes, then add 30 µℓ of TNBS solution. Cap, vortex, and place the tubes in a 70°C heating block for 30 min.

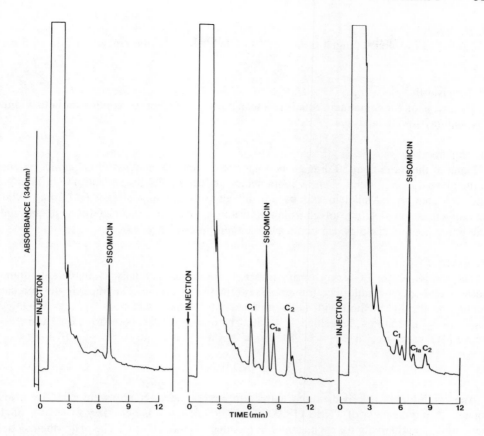

FIGURE 1. Left: Chromatograms of a drug-free serum; middle: chromatogram of a patient's serum containing 8.5 µg/mℓ of gentamicin. right: chromatogram of a patient's serum containing 1.6 µg/mℓ of gentamicin.

Transfer each sample into a labeled 12 mℓ glass centrifuge tube, add 2 mℓ of chloroform and shake for 3 min. Centrifuge the samples at 2000 rpm (210 × g) for 4 min, aspirate the top layer, and evaporate the chloroform layer in an Evapo-Mix® at 50°C. Reconstitute the sample with 100 µℓ of acetonitrile, inject 50 µℓ of sample onto the chromatograph and elute with the mobile phase at a flow rate of 2.0 mℓ/min. The column is maintained at 50°C, and the column effluent is monitored at 340 nm (Figure 1).

### E. Calculations

1. Serum standards are carried through the procedure and used to prepare a standard curve based on relative peak heights or peak area ratios of gentamicin $C_1$, $C_{1a}$, and $C_2$ isomers.
2. Calculate the concentration of gentamicin isomers in the specimen by direct comparison with the data obtained from the standard curve of the serum standards. The total concentration of gentamicin is calculated by adding the concentrations of each isomer.

## III. RESULTS

### A. Linearity[1]

Drug concentrations and peak area ratios are linear to 50 mg/ℓ. The lower limit of sensitivity is approximately 0.5 mg/ℓ.

## B. Recovery[1]
Analytical recoveries ranged from 97 to 99% in a concentration range of 1.5 to 15.5 mg/$\ell$.

## C. Precision[1]
The precision for gentamicin equals less than 3.5% coefficient of variation for both within-day and day-to-day analyses.

## D. Interference[1]
None of the more than 30 drugs tested interfered with the analyses of gentamicin. Other aminoglycoside antibiotics such as tobramycin, amikacin, and kanamycin do not interfere with the analyses. Amikacin and kanamycin elute in the solvent font (<1 min), while tobramycin is completely resolved from the various gentamicin isomers. Grossly hemolyzed, icteric, or lipemic samples can be assayed without interference.

## E. Accuracy
The results of our LC assay were compared with a commercially available radioimmunoassay and enzyme multiplied immunoassay (EMIT) technique. The linear regression data comparing the LC method with radioimmunoassay method were n = 57, r = 0.995, y-intercept = 0.025 mg/$\ell$, and slope = 4.959. The regression data comparing the LC method with EMIT were: n = 33, r = 0.981, slope = 0.967, and y-intercept = 0.074 mg/$\ell$.[1]

## IV. COMMENTS

The conditions for derivatization, extraction, and chromatography were carefully investigated. A large excess of TNBS (18,000 to 20,000 $M$ ratio) is necessary to yield a single derivative quantitatively for each isomer in less than 30 min at 70°C. The pH of the reaction mixture is very critical for complete derivatization. The optimum pH for this reaction is between 9.5 to 10. The nonpolar gentamicin derivatives were extracted from the crude mixture into chloroform. This extraction procedure extended the useful life of the analytical column.

## REFERENCE

1. **Kabra, P. M., Bhatnagar, P. K., Nelson, M. A., and Marton, L. J.**, Liquid chromatographic determination of gentamicin in serum with spectrophotometric detection, *J. Anal. Toxicol.*, submitted.

Chapter 7

# SERUM TOBRAMYCIN WITH SPECTROPHOTOMETRIC DETECTION

### Pokar M. Kabra

## TABLE OF CONTENTS

I. Introduction ................................................................. 34

II. Principle ..................................................................... 34

III. Materials and Methods ................................................. 34
    A. Equipment ............................................................ 34
    B. Reagent ................................................................ 34
    C. Standards ............................................................. 35
    D. Procedure ............................................................. 35
    E. Calculation ........................................................... 35

IV. Results ....................................................................... 36
    A. Linearity .............................................................. 36
    B. Recovery .............................................................. 36
    C. Precision .............................................................. 36
    D. Interference .......................................................... 36
    E. Comparison with Immunoassays ............................. 37

V. Comments .................................................................. 37

References ........................................................................ 37

## I. INTRODUCTION

Tobramycin, an aminoglycoside antibiotic, is widely used against Gram-negative bacterial infections. It has a narrow therapeutic range and can cause severe nephro- and ototoxicity at serum concentrations exceeding 12 mg/$\ell$.[1] Therefore, it is essential to monitor serum levels during therapy and maintain them within the prescribed therapeutic range for optimum efficacy. Methods for measuring serum tobramycin concentration include microbiological, radioenzymatic, and immunoassays.[2,3] These methods suffer from several deficiencies such as slow turn around time and limited specificity due to interference by other antimicrobial agents. Several liquid chromatographic (LC) procedures have been reported for the measurement of aminoglycoside antibiotics.[4-7] Most of these methods require either precolumn or postcolumn derivatization for fluorescence detection. A nonfluorescent method which uses simple LC equipment is preferable to more expensive and more complicated pre- and postcolumn LC-fluorescence methods.

## II. PRINCIPLE

Tobramycin and added internal standard, sisomicin, are derivatized at alkaline pH with 2,4,6 trinitrobenzene sulfonic acid (TNBS) following serum protein precipitation. The derivatives are adsorbed from serum on a Bond-Elut® extraction column. After the polar reaction products are washed from the column with buffered methanol, tobramycin and internal standard derivatives are eluted with acetonitrile and an aliquot of the eluate is injected onto a reversed phase octyl column and separated using an acetonitrile/phosphate buffer mobile phase. The derivatives are detected by their absorption at 340 nm and quantitated from either their peak height or peak area ratios.

## III. MATERIALS AND METHODS

### A. Equipment

Liquid chromatographic systems equivalent to the following are recommended: a Model Series 2 or Series 3 liquid chromatograph equipped with a Model LC-100 column oven, a Model LC-75 variable wavelength detector, and a Sigma® 10 data system (all from Perkin-Elmer Corp., Norwalk, Conn.) is used. The reversed phase column 25 cm × 4.6 mm (Ultrasphere®-octyl, Altex Scientific, Berkeley, Calif.) is mounted in the oven. The sample is injected into a Model 7105 valve (Rheodyne, Cotati, Calif.) mounted on the chromatograph. A Model 5412 Eppendorf® centrifuge with 1.5 m$\ell$ polypropylene tubes (Brinkman Instruments, Inc., Westbury, N.Y.) are used.

### B. Reagent

All chemicals used are of reagent grade. Acetonitrile, and methanol, all distilled in glass (Burdick and Jackson Laboratories, Inc., Muskegon, Mich.). Phosphate buffer: 50 mmol/$\ell$ is prepared by dissolving 6.8 g of $KH_2PO_4$ in 1 $\ell$ of distilled water. Mobile phase: this is a solution of 700 m$\ell$ of acetonitrile in 300 m$\ell$ of 50 mmol/$\ell$ phosphate buffer. The mobile phase is titrated with phosphoric acid to pH 3.5. Tris buffer, 2 mol/$\ell$, pH 10.3, is prepared by dissolving 24.2 g of Trizma® base (Sigma Chemical Co., St. Louis) in 100 m$\ell$ of distilled water. This solution is stable for at least 1 year at 4°C. 2,4,6, Trinitrobenzene 1-sulfonic acid (TNBS, Sigma), 250 g/$\ell$, is prepared by dissolving 2.5 g of trinitrobenzene sulfonic acid in 10 m$\ell$ of distilled water. This solution is stable for 1 month at 4°C. Stock washbuffer, 1 mol/$\ell$ $K_2HPO_4$, is prepared by dissolving 87 g of $K_2HPO_4$ in 500 m$\ell$ of distilled water. Washing wash solution (methanol/phosphate buffer 0.1 mol/$\ell$ $K_2HPO_4$, pH 8.5, 50/50 by vol) is prepared by transferring 10 m$\ell$ of stock wash buffer into a 250 m$\ell$ graduated cylinder and adding 90 mℓ of distilled water and 100 mℓ of methanol. Adjust the pH to 8.5 with phosphoric acid. This solution is stable for 1 year at ambient temperature.

Vac-Elut® vacuum chamber and Bond-Elut® C-18 extraction columns can be obtained from Analytichem International, Inc., Harbor City, Calif. Polypropylene tubes, 1.5 mℓ capacity, and an Eppendorf® Model 5412 centrifuge are from Brinkman Instruments, Inc., Westbury, N.Y.

### C. Standards

Tobramycin can be obtained from Eli Lilly and Co., Indianapolis, Ind. and sisomicin from Schering Corp., Bloomfield, N.J. The tobramycin stock standard is prepared by dissolving 25 mg in 100 mℓ of water. The solution is stable at 4°C for at least 6 months. The stock internal standard is prepared by dissolving 25 mg of sisomicin in 100 mℓ of acetonitrile. The solution is stable at 4°C for at least 6 months. The working internal standard is prepared by diluting the stock internal standard 25-fold with acetonitrile. The solution is stable for 1 month at 4°C. The working tobramycin serum standards (5 and 10 mg/ℓ) are prepared by diluting 200 and 400 μℓ of stock standard with 9.8 and 9.6 mℓ drug-free serum respectively. The serum standards are stable for at least 1 month at 4°C.

### D. Procedure

To 50 μℓ of serum sample, serum standard, or control in a 1.5 mℓ polypropylene tube, add 25 μℓ of 2 mol/ℓ tris-buffer and 100 μℓ of working internal standard (sisomicin, 10 mg/ℓ). Vortex-mix and centrifuge all tubes for 1 min in Eppendorf® centrifuge at 15,000 g. Decant the supernate into second set of appropriately labeled polypropylene tubes, then add 20 μℓ of TNBS solution. Cap, vortex, and place the tubes in 70°C heating block for 30 min. For each sample, place a Bond-Elut® $C_{18}$ extraction column on the top of the Vac-Elut® chamber, connect the vacuum to the chamber. Pass two column volumes of methanol followed by two column volumes of water through each column. Disconnect vacuum and fill each column with 700 μℓ of working wash solution, followed by approximately 200 μℓ of derivatized sample. Connect vacuum to the chamber and pass three volumes of working wash solution through each column. Disconnect vacuum and place a rack containing labeled 10 × 75mm glass tubes in the Vac-Elut® chamber. The position of each glass tube should match the corresponding Bond-Elut® column. pipette 300 μℓ of acetonitrile onto each column and connect vacuum. Remove the rack of tubes from the vacuum chamber, shake in tubes to mix the eluate, and inject approximately 50 μℓ of each eluate onto the chromatograph. Eluate with the mobile phase at a flow rate of 3.0 mℓ/min. The column temperature is maintained at 50°C and the column effluent is monitored at 340 nm (Figure 1).*

### E. Calculation

1. Serum standards are carried through the procedure and used to prepare a standard curve based on relative peak height or peak area ratios.
2. Calculate the concentration of tobramycin in the specimen by direct comparison with the data obtained from the standard curve of serum standards.

---

\* The optimized conditions for derivatization, extraction, and chromatography were carefully investigated. A large excess of TNBS (18,000 to 20,000 $M$ ratio) was necessary to yield a single tobramycin derivative in <30 min quantitative. At 70°C the derivatization was complete in <30 min. The pH of the reaction mixture was very critical for complete derivatization. The optimum pH for this reaction was found to be between 9.5 to 10.0. The Bond-Elut® $C_{18}$ reversed-phase columns were used to extract the tobramycin and sisomicin derivatives from the crude mixture. The solid phase extraction procedure simplified sample preparation and eliminated the large solvent front. Chromatographic time was <4.5 min.

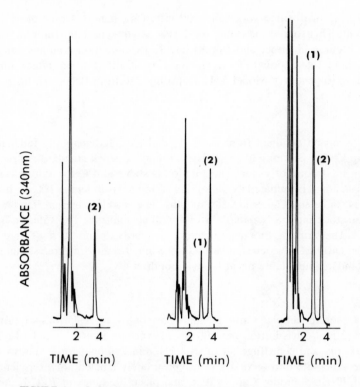

FIGURE 1. Left: chromatogram of a tobramycin-free serum; middle: chromatogram of a patient's serum containing 2.3 mg/ℓ of tobramycin; right: chromatogram of a patient's serum containing 7.5 mg/ℓ of tobramycin. Peak (1) represents tobramycin; peak (2) represents internal standard (sisomicin). (From Kabra, P. M., Bhatnagar, P. K., Nelson, M. A., Wall, J. N., and Marton, L. J., *Clin. Chem.*, 29, 672, 1983. With permission.)

## IV. RESULTS

### A. Linearity[8]

Drug concentrations and peak area ratios are linearily related up to 50 mg/ℓ with a lower limit of sensitivity of approximately 0.2 µg/ℓ.

### B. Recovery[8]

Analytical recoveries ranged from 94 to 98.6% in a concentration range of 1 to 25 mg/ℓ.

### C. Precision[8]

The precision for tobramycin equals less than 5% coefficient of variation for both within-day and day-to-day analyses.

### D. Interference[8]

None of more than 30 drugs tested interfered with the analyses of tobramycin. Other aminoglycoside antibiotics such as gentamicin, amikacin, and kanamycin, do not interfere with the analysis. Amikacin and kanamycin elute in the solvent front (<1 min), while gentamicin isomers are completely resolved from tobramycin. Grossly hemolyzed, icteric, or lipemic samples can be assayed without interference.

## E. Comparison with Immunoassays

The liquid chromatographic method correlated well with an established radioimmunoassay and enzyme multiplied immunoassay. The correlation coefficients were 0.968 and 0.981, respectively, for the two methods.

## V. COMMENTS

The pH of the derivatizing mixture should be above 10. Tobramycin and internal standard are not quantitatively derivatized below pH 10. The pH of the working wash buffer should be 8.5. If the pH of the buffer is below pH 8.0, the internal standard is not retained on the extraction column. Above pH 9.0 tobramycin is not retained on the extraction column.

## REFERENCES

1. **Anhalt, J. P.**, Antibiotics, in *Liquid Chromatography in Clinical Analysis*, Kabra, P. M. and Marton, L. J., Eds., Humana Press, Clifton, N.J., 1981, 163.
2. **Stevens, P., Young, L. S., and Hewitt, W. L.**, Radioimmunoassay, acetylating radioenzymatic assay, and microbioassay of gentamicin: a comparative study, *J. Lab. Clin. Med.*, 86, 349, 1975.
3. **Maitra, S. K., Yoshikawa, T. T., Guze, L. B., et al.**, Determination of aminoglycoside antibiotics in biological fluids. A review, *Clin. Chem.*, 25, 1361, 1979.
4. **Peng, G. W., Gadalle, M. A. F., Peng, A., et al.**, High-pressure liquid-chromatographic method for determination of gentamicin in plasma, *Clin. Chem.*, 23, 1838, 1977.
5. **Anhalt, J. P. and Brown, S. D.**, High-performance liquid chromatographic assay of aminoglycoside antibiotics in serum, *Clin. Chem.*, 24, 1940, 1978.
6. **Maitra, S. K., Yoshikawa, T. T., Hansen, J. L., et al.**, Serum gentamicin assay by high performance liquid chromatography, *Clin. Chem.*, 23, 2275, 1977.
7. **Back, S. E., Nilssen-Ehle, I., and Nilseen-Ehle, P.**, Chemical assay, involving liquid chromatography, for aminoglycoside antibiotics in serum, *Clin. Chem.*, 25, 1222, 1979.
8. **Kabra, P. M., Bhatnagar, P. K., Nelson, M. A., Wall, J. N., and Marton, L. J.**, Liquid chromatographic determination of tobramycin in serum with spectrophotometric detection, *Clin. Chem.*, 29, 672, 1983.

Chapter 8

# ANTIDYSRHYTHMIC DRUGS BY ULTRAVIOLET/FLUORESCENCE DETECTION

**Pokar M. Kabra**

## TABLE OF CONTENTS

I. Introduction ................................................................. 40

II. Principle ..................................................................... 40

III. Materials and Methods ................................................. 40
    A. Equipment ............................................................. 40
    B. Reagent ................................................................ 40
    C. Standards .............................................................. 41
    D. Procedure (UV Detection) ...................................... 42
    E. Calculations .......................................................... 44

IV. Results ...................................................................... 44
    A. Linearity ............................................................... 44
    B. Recovery .............................................................. 44
    C. Precision .............................................................. 44
    D. Sensitivity ............................................................. 45
    E. Interference .......................................................... 45
    F. Accuracy .............................................................. 45

V. Comments ................................................................. 45

References ......................................................................... 45

## I. INTRODUCTION

The antidysrhythmic agents (procainamide, lidocaine, quinidine, disopyramide, and propranolol) are widely used cardiac agents. They have narrow therapeutic ranges and adverse side effects. A good correlation between serum concentrations and clinical effects is obtained for these drugs.[1] Therefore, it is desirable to monitor their serum concentrations as a therapeutic guide. N-Acetylprocainamide (NAPA) is a major metabolite of procainamide and it possesses comparable toxicity as well as therapeutic potency to procainamide. The rate of metabolism is usually genetically determined.[2]

The antidysrhythmic drugs and their bioactive metabolites have been determined by several methods, including colorimetry, fluorometry, gas chromatography, UV spectroscopy, immunoassay, and liquid chromatography.[3] Liquid chromatography (LC) is well-suited for the simultaneous analysis of various parent drug-metabolite combinations in serum. We report a method for the simultaneous analysis of procainamide, NAPA, lidocaine, quinidine, disopyramide, N-desisopropyl disopyramide, and propranolol by liquid chromatography.[4]

## II. PRINCIPLE

The antidysrhythmic drugs and the internal standard are extracted from 200 to 500 µℓ of serum at alkaline pH into methylene chloride or ether. The organic layer is separated, evaporated to dryness, and reconstituted in mobile phase. An aliquot of the sample is injected onto a reversed-phase octyl column and the drugs are eluted with an acetonitrile-phosphate buffer mobile phase. The drugs are detected by either UV or fluorescence spectrophotometry and quantitated from either their peak heights or peak areas.

## III. MATERIALS AND METHODS

### A. Equipment

Liquid chromatograph systems equivalent to the following are recommended. A Model Series 2 or Series 3 liquid chromatograph equipped with a 7105 injection valve, a Model 75 or 65T variable wavelength detector, a Model 6510 LC fluorescence detector, a Model LC 100 temperature controlled oven (all from Perkin-Elmer Corp., Norwalk, Conn.), a reversed-phase column Ultrasphere® octyl 5 µm 25 cm × 4.6 mm (Altex Scientific, Berkeley, Calif.) mounted in the oven, and a Sigma® 10 data system (Perkin-Elmer) is used.

### B. Reagent

All chemicals used are of reagent grade. Acetonitrile, propanol-2, and methylene chloride, all distilled in glass, can be obtained from Burdick and Jackson Laboratories, Inc., Muskegon, Mich. Ethyl ether anhydrous, analytical reagent grade, is from Mallinckrodt, Inc., St. Louis. Carbonate buffer, 1 mol/ℓ, pH 10.8, is prepared by dissolving 10.6 g of sodium carbonate in 100 mℓ of water, and pH is adjusted to 10.8 with 1 mol/ℓ sodium bicarbonate.

*Mobile phase buffers:*

1. Phosphate buffer, 25 mmol/ℓ, pH 3.0, is prepared by dissolving 3.4 g of $KH_2PO_4$ in water. The pH of the solution is adjusted to 3.0 with $H_3PO_4$ and diluted to 1 ℓ with water.
2. Phosphate buffer, 75 mmol/ℓ, pH 3.4, is prepared by dissolving 10.2 g of $KH_2PO_4$ in water. The pH of the solution is adjusted to 3.4 with $H_3PO_4$ and diluted to 1 ℓ with water.

3. Phosphate buffer, 100 mmol/ℓ, pH 3.0, is prepared by dissolving 13.6 g of $KH_2PO_4$ in water. The pH is adjusted to 3.0 with $H_3PO_4$, and diluted to 1 ℓ with water.

*Mobile phase for a gradient analysis:*

1. Acetonitrile, 7% in 25 mmol/ℓ, phosphate buffer, (pH 3.0) is prepared by diluting 70 mℓ of acetonitrile and 930 mℓ of 25 mmol/ℓ phosphate buffer.
2. Acetonitrile, 30% in 100 mmol/ℓ, phosphate buffer, (pH 3.0) is prepared by diluting 300 mℓ of acetonitrile with 700 mℓ of 100 mmol/ℓ phosphate buffer.

*Mobile phase for isocratic analysis:* acetonitrile, 25.5% in 75 mmol/ℓ, phosphate buffer, (pH 3.4) is prepared by diluting 255 mℓ of acetonitrile with 745 mℓ of 75 mmol/ℓ phosphate buffer.

**C. Standards**

Procainamide HCl can be obtained from Pfaltz and Bauer, Inc., Flushing, N.Y.; *N*-acetylprocainamide and *N*-propionyl procainamide from Pierce Chemical Co., Rockford, Ill.; quinidine HCl from K & K Laboratories, Inc., Plainview, N.Y.; lidocaine HCl monohydrate and lidocaine metabolites from Astra Pharmaceutical Products, Inc., Worcester, Mass.; disopyramide, *N*-desisopropyl disopyramide, and *P*-chlorodisopyramide from G. D. Searle & Co., Chicago; propranolol HCl and pronethalol HCl from Ayerst Laboratories, Inc., New York.

A reference standard for procainamide and *N*-acetylprocainamide is prepared by dissolving 11.55 mg of procainamide HCl, 10 mg of *N*-acetylprocainamide, and 10 mg of *N*-propionyl procainamide in 100 mℓ of methanol. This reference standard is stable for 6 months at 4°C.

A stock internal standard for procainamide and NAPA is prepared by dissolving 10 mg of *N*-propionyl procainamide in 100 mℓ of methanol. This standard is stable for 6 months at 4°C.

A working internal standard is prepared by diluting the stock internal standard tenfold with water. The working internal standard is stable for 3 months at 4°C.

A reference standard for lidocaine, quinidine, disopyramide, *N*-desisopropyl disopyramide, and propranolol is prepared by dissolving 10 mg each of disopyramide, *N*-desisopropyl disopyramide; 12.33 mg of lidocaine HCl.$H_2O$; 11.70 mg of quinidine HCl; 1.1 mg of propranolol HCl, and 10 mg of *P*-chlorodisopyramide in 100 mℓ of 10% methanol. This solution is stable for 6 months at 4°C.

A stock internal standard is prepared by dissolving 10 mg of *p*-chlorodisopyramide in 100 mℓ of water. This solution is stable for 6 months at 4°C.

A working internal standard is prepared by diluting the stock internal standard tenfold with water for the analysis of quinidine, disopyramide, and lidocaine. For propranolol analysis, the stock internal standard is diluted 100-fold with water. Both of these working internal standards are stable for 1 month at 4°C.

The reference standard for quinidine and propranolol analysis by fluorescence detection is prepared by dissolving 5.85 mg of quinidine HCl and 5 mg of pronethalol HCl in 1 ℓ of water. For the propranolol standard, dissolve 5.7 mg of propranolol HCl and 50 mg of pronethalol HCl in 1 ℓ of water.

The stock internal standard for quinidine and propranolol by fluorescence detection is prepared by dissolving 10 mg of pronethalol HCl in 100 mℓ of 0.1 mol/ℓ HCl. This solution is stable for 3 months at 4°C.

The working internal standard for fluorescence detection is prepared by diluting 1 mℓ of concentrated internal standard 40-fold with water for quinidine analysis and 400-fold for propranolol analysis.

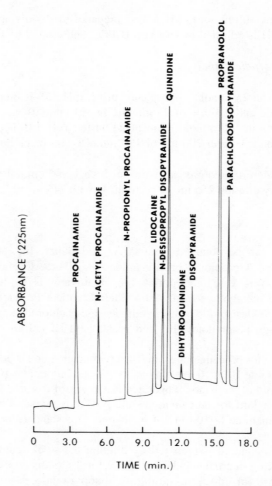

FIGURE 1. Chromatogram of a standard mixture of drugs run under gradient conditions. (From Kabra, P. M., Chen, S. W., and Marton, L. J., *Ther. Drug. Monit.*, 3, 91, 1981. With permission.)

## D. Procedure (UV Detection)

Place 500 μℓ of standard, serum sample, and control into 12 mℓ glass centrifuge tubes (use 1 mℓ for propranolol). To each tube add 100 μℓ of 1 mol/ℓ NaOH and 500 μℓ of working internal standard. Vortex-mix each tube for 10 sec, then add 6 mℓ of dichloromethane. Shake the tubes for 5 min and then centrifuge for 5 min at 2000 rpm (210 × g) to separate the phases. Aspirate off the upper aqueous and protein layers and pour the dichloromethane layer into a clean 12 mℓ glass centrifuge tube (use polypropylene tubes for quinidine analysis to avoid glass adsorption). Evaporate the organic layer to dryness at 50°C under a gentle stream of air. Dissolve the residue in 50 μℓ of methanol and inject 20 μℓ onto the chromatograph (inject the entire sample for propranolol). The reversed-phase octyl column is eluted with an initial mobile phase containing 7% acetonitrile in 25 mmol/ℓ phosphate buffer and increasing the concentration to 30% acetonitrile in 100 mmol/ℓ phosphate buffer in 10 min (linear gradient). The flow rate is set at 2.0 mℓ/min at 50°C and the column effluent is monitored at 225 mm (Figure 1). For isocratic analysis of lidocaine, disopyramide, quinidine, and propranolol, the reversed-phase octyl column is eluted with a mobile phase containing 25.5% acetonitrile in 75 mmol/ℓ phosphate buffer (pH 3.4) at a

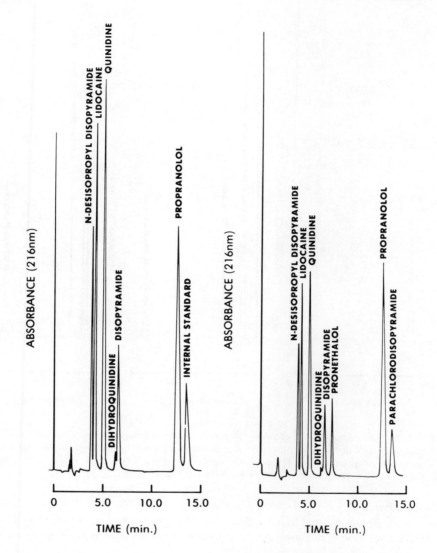

FIGURE 2. Left: chromatogram of a standard mixture of drugs run under isocratic condition; right: chromatogram of a standard mixture of drugs with pronethalol (internal standard for fluorescence detection). (From Kabra, P. M., Chen, S. W., and Marton, L. J., *Ther. Drug. Monit.*, 3, 91, 1981. With permission.)

flow rate of 1.5 min at 40°C. The column effluent is monitored at 216 nm (Figures 2 and 3).

Procedure for quinidine and propranolol (fluorescence detection): place 200 µℓ of serum sample, standard, or control into 12 mℓ glass centrifuge tubes. To each tube add 100 µℓ of 1 mol/ℓ carbonate buffer (pH 10.8) and 200 µℓ of working internal standard. Vortex mix for 10 sec, centrifuge briefly at 2500 rpm (210 × g), and freeze the aqueous layer (bottom layer) in an acetone dry ice bath. Decant ether into a clean 12-mℓ glass tube (polypropylene tube for quinidine analysis) and evaporate at 50°C under a stream of air. Dissolve the residue in 100 µℓ of methanol and inject 20 µℓ onto the chromatograph. The reversed-phase octyl column is eluted with a mobile phase containing 25.5% acetonitrile in 75 mmol/ℓ phosphate buffer (pH 3.4) at a flow rate of 1.5 mℓ/min at 40°C. The fluorescence detector is set to read excitation wavelength at 290 nm with an emission wavelength of 350 nm.

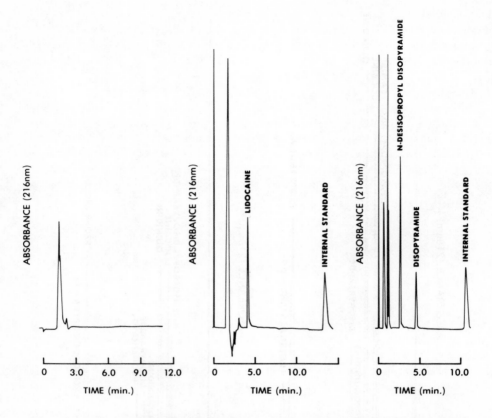

FIGURE 3. Left: chromatogram of a drug-free serum; middle: chromatogram from a patient's serum containing 4.7 mg/ℓ of lidocaine; right: chromatogram from a patient's serum containing 9.8 mg/ℓ of N-desisopropyl disopyramide, and 6.3 mg/ℓ of disopyramide. (From Kabra, P. M., Chen, S. W., and Marton, L. J., *Ther. Drug. Monit.*, 3, 91, 1981. With permission.)

### E. Calculations

Construct a calibration curve for each drug, using the analyte/internal standard peak height ratio. The concentration of unknown drug is calculated by direct comparison with the standard curve.

## IV. RESULTS

### A. Linearity

Peak height ratios of procainamide, N-acetylprocainamide, disopyramide, N-desisopropyldisopyramide, and quinidine to internal standard are linearly related from to at least 20 μg/mℓ for each drug. Lidocaine and proporanolol are linearly related from 1 to 10 μg/mℓ and 20 to 200 ng/mℓ, respectively.

### B. Recovery

Analytical recoveries for all of these drugs ranges from 79 to 112% in the subtherapeutic to toxic concentration range.[4]

### C. Precision

The precision for all antidysrhythmic drugs equals less than 11.5% coefficient of variation for both within-day and day-to-day analyses.[4]

## D. Sensitivity

The lower limit of detection for each drug except for propranolol is 0.1 mg/$\ell$. For propranolol the lowest limit of detection is 10 and 1 μg/$\ell$, respectively, with UV and fluorescence detection.[4]

## E. Interference

Diazepam and flurazepam interfere with quinidine analysis by UV detection but do not interfere with its fluorescence detection. Meperidine elutes close to disopyramide and might interfere with its analysis when present at toxic concentrations. A recently isolated metabolite of quinidine (N-oxide metabolite) coelutes with quinidine under the assay conditions. However, this metabolite can be separated from quinidine by employing a 24% acetonitrile in phosphate buffer mobile phase (pH 2.8). Lipemic, icteric, or hemolyzed sera do not interfere with the analysis.

## F. Accuracy

The method correlated well with gas-liquid chromatographic, liquid chromatographic, and immunoassay methods. The correlation coefficient ranged from 0.962 to 0.997.

## V. COMMENTS

The reversed-phase octyl column eliminated the poor resolution and peak tailing associated with an octadecyl column. Using an octyl column, ion pairing, or ion suppression were not necessary. A composition of 25.5/74.5 (acetonitrile/phosphate buffer) proved to be the best mobile phase for the isocratic analysis of lidocaine, quinidine, disopyramide, and propranolol. Gradient chromatography employing acetonitrile and phosphate buffer provided baseline resolution for all seven antidysrhythmic agents. Detection at 216 nm provided adequate sensitivity for the therapeutic monitoring of propranolol and lidocaine. Fluorescence detection provided even better sensitivity for quinidine and propranolol and was especially useful for microsamples. The occasional variation in the recovery of quinidine with the use of glass centrifuge tubes was alleviated by substituting polypropylene tubes during the final evaporation step. Usually the concentration of the quinidine metabolite (previously mentioned) is insignificant and the modified procedure described is not warranted.

## REFERENCES

1. **Kalman, S. M. and Clark, D. R.**, Antiarrhythmic drugs, in *Drug Assay: The Strategy of Therapeutic Drug Monitoring*, Masson Publishing, New York, 1979, 31.
2. **Harris, H.**, *The Principles of Human Biochemical Genetics*, 2nd ed., Elsevier, New York, 1975, 254.
3. **Vandemark, F. L.**, Theophylline and antiarrhythmics, in *Liquid Chromatography in Clinical Analysis*, Kabra, P. M. and Marton, L. J., Eds., Humana Press, Clifton, N.J., 1981, 139.
4. **Kabra, P. M., Chen, S. W., and Marton, L. J.**, Liquid-chromatographic determination of antidysrhythmic drugs: procainamide, lidocaine, quinidine, disopyramide, and propranolol, *Ther. Drug. Monit.*, 3, 91, 1981.

Chapter 9

# PROCAINAMIDE AND *N*-ACETYL PROCAINAMIDE BY ULTRAVIOLET DETECTION

### George R. Gotelli and Jeffrey H. Wall

## TABLE OF CONTENTS

| | | |
|---|---|---|
| I. | Introduction | 48 |
| II. | Principle | 48 |
| III. | Materials and Methods | 48 |
| | A. Equipment | 48 |
| | B. Reagents | 48 |
| | C. Standards | 48 |
| | D. Procedure | 48 |
| | E. Calculations | 49 |
| IV. | Results | 50 |
| | A. Optimization of Chromatography | 50 |
| | B. Linearity | 50 |
| | C. Recovery | 50 |
| | D. Reproducibility | 50 |
| | E. Interferences | 50 |
| | F. Accuracy | 50 |
| V. | Comments | 50 |
| References | | 51 |

## I. INTRODUCTION

Procainamide is used in the treatment of cardiac arrhythmias. This amide of procaine is hydrolyzed slowly in vivo and thus procainamide plasma levels decline less rapidly than procaine. The plasma concentration of procainamide is frequently measured as an assessment of therapeutic response and to avoid toxicity.

Colorimetric,[1] fluorometric,[2] and gas-liquid chromatographic[3] methods have been reported for the analysis of procainamide. Many of these methods, however, either do not measure the active metabolite of procainamide, N-acetyl procainamide, or both the metabolite and parent drug are measured as one. The described liquid chromatographic (LC) method, modified from Kabra et. al.,[4] is simple, rapid, specific, and quantitates both the parent drug and metabolite (Figure 1).

## II. PRINCIPLE

Procainamide, N-acetyl procainamide, and added internal standard are extracted from alkalinized serum with methylene chloride. The drugs are back-extracted into phosphoric acid and an aliquot of the phosphoric acid is injected onto a reversed-phase octyl column. The drugs are eluted with mobile phase, detected at 280 nm, and quantitated by peak height ratios.

## III. MATERIALS AND METHODS

### A. Equipment

A chromatography system equivalent to the following is recommended. A series 1 pump, an LC-100 oven, and an LC-15 detector with a 280 nm filter (all from Perkin-Elmer Corp., Norwalk, Conn.). Also a Rheodyne® 7105 injection valve (Rheodyne, Cotati, Calif.), an Altex Ultrasphere® Octyl column, 15 cm × 4.6 mm i.d. (Altex Scientific, Berkeley, Calif.), and a 10 mV strip-chart recorder.

### B. Reagents

Sodium hydroxide (1 mol/ℓ), reagent grade. Phosphoric acid (0.043 $N$) reagent grade, equivalent to a 0.1% solution. Potassium dihydrogen phosphate buffer (1 mol/ℓ, pH 4.4). Mobile phase — to a 2-ℓ flask add 240 mℓ methanol, 50 mℓ tetrahydrofuran, and 20 mℓ of phosphate buffer (pH 4.4), add distilled water to volume.

### C. Standards

Prepare stock 100 mg/dℓ procainamide and N-acetyl procainamide in 10% methanol. The procainamide was purchased from Pfaltz and Bauer, Inc, Flushing, N.Y. and N-acetyl procainamide from Pierce Chemical Co., Rockford, Ill.

Prepare a working serum standard of 10 μg/mℓ of procainamide and N-acetyl procainamide in serum by adding an appropriate amount of the stock standard to drug-free serum. This serum standard is stable for 2 weeks at 4°C.

Prepare a stock internal standard of 10 mg of N-propionyl procainamide (Pierce Chemical Co.) in 100 mℓ of methanol.

Prepare a working internal standard by diluting the stock internal standard with water to give a final concentration of 10 μg/mℓ. The working internal standard is stable at least 6 months at 4°C.

### D. Procedure

Mix 100 μℓ of working serum standard or unknown serum with 100 μℓ of the working

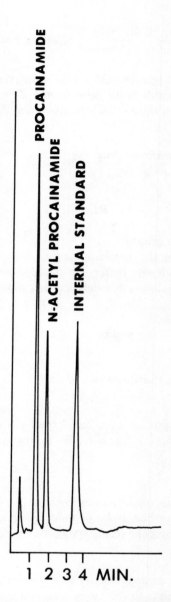

FIGURE 1. Sample containing 5.4 μg of procainamide and 3.4 μg of N-acetyl procainamide per milliliter of serum (obtained in the authors' laboratory)

internal standard. Add 1 drop of 1 mol/ℓ sodium hydroxide and 1 mℓ of methylene chloride to each tube. Vortex or shake for 1 min. Centrifuge briefly to separate layers and aspirate to waste the upper aqueous layer. Decant the methylene chloride into clean tubes and add 200 μℓ of 0.1% phosphoric acid. Vortex or shake 1 min. Allow 5 min for phase separation, then inject 20 μℓ of the upper phosphoric acid layer onto the column. HPLC conditions are flow rate, 3 mℓ/min; oven, 50°C; and detector, 280 nm.

## E. Calculations

Calculate a response factor (RF) from the working serum standard chromatogram as follows:

$$\frac{\text{peak height of Int. Std.}}{\text{peak height of procainamide}} = \text{RF for procainamide}$$

Calculate an RF for N-acetyl procainamide in the same manner by substituting the peak height of the N-acetyl procainamide in the denominator of the fraction.

Calculate the concentration of drug in each sample from their respective chromatograms and RF's as follows:

$$\frac{\text{peak height of unknown drug}}{\text{peak height of Int. Std.}} \times \text{RF} \times 10 = \text{conc. of unknown drug } (\mu g/m\ell).$$

## IV. RESULTS

### A. Optimization of Chromatography

The use of tetrahydrofuran in the mobile phase improves peak resolution and symmetry without the use of high ionic strength buffers which are sometimes troublesome. The 50°C oven temperature improves resolution and stabilizes retention times.

### B. Linearity

Peak height ratios and drug concentration are linear from 0 to 100 $\mu g/m\ell$.

### C. Recovery

Both absolute and analytical recoveries exceed 95%.

### D. Reproducibility

Within run CV is 2.9% for procainamide and 2.6% for N-acetyl procainamide. The CV for daily analysis of controls over 1 year is 6.3% for procainamide and 6.0% for N-acetyl procainamide.

### E. Interferences

Common anticonvulsants, analgesics, antiarrhythmics, tricyclics, and benzodiazepines were tested for possible interference. None of those tested were found to interfere. Daily experience with this method for over one year has produced very few unidentified interferences.

### F. Accuracy

The method has not been compared directly to a specific preexisting method; however, comparison of the assayed results to the target values of numerous College of American Pathologists (CAP) and Center for Disease Control (CDC) survey samples yielded the following: procainamide r = 0.987, slope = 0.932, y-intercept = 0.47, n = 8. N-acetyl procainamide r = 0.994, slope = 1.073, y-intercept = 0.8, n = 8.

## V. COMMENTS

Back extraction of these basic drugs into phosphoric acid eliminates time consuming evaporative steps,[4] avoids problems of glass adsorption, and eliminates interferences. The solution being injected onto the column is very clean, and column life should exceed 1 year.

# REFERENCES

1. **Sitar, D. S., Graham, D. N., Rango, R. E., Dusfresne, L. R., and Ogilvie, R. I.,** Modified colorimetric methods for procainamide in plasma, *Clin. Chem.,* 22, 379, 1976.
2. **Koch-Weser, J. and Klein, S. W.,** Procainamide dosage schedules, plasma concentrations and clinical effects, *JAMA,* 215, 1454, 1971.
3. **Drayer, D. E., Reidenberg, M. M., and Sevy, R. W.,** N-acetyl procainamide: an active metabolite of procainamide, *Proc. Soc. Exp. Biol. Med.,* 146, 358, 1974.
4. **Kabra, P. M., Chen, S. H., and Marton, L. J.,** Liquid-chromatographic determination of antidysrhythmic drugs: procainamide, lidocaine, quinidine, disopyramide, and propranol, *Ther. Drug Monit.,* 3, 91, 1981.

Chapter 10

## PROPRANOLOL BY FLUORESCENCE DETECTION

**George R. Gotelli and Jeffrey H. Wall**

## TABLE OF CONTENTS

I. Introduction .................................................................. 54

II. Principle ..................................................................... 54

III. Materials and Methods ....................................................... 54
    A. Equipment ............................................................... 54
    B. Reagents ................................................................ 54
    C. Standards ............................................................... 54
    D. Procedure ............................................................... 54
    E. Calculations ............................................................ 55

IV. Results ...................................................................... 55
    A. Optimization of Chromatography ........................................ 55
    B. Linearity ............................................................... 55
    C. Recovery ................................................................ 56
    D. Reproducibility ......................................................... 56
    E. Interference ............................................................ 56
    F. Accuracy ................................................................ 56

Reference ......................................................................... 56

## I. INTRODUCTION

Propranolol is a beta-adrenergic blocking agent which causes bradycardia, prolonged systole, and a decrease in blood pressure. It is used most commonly in the management of hypertension and its blood concentration is frequently determined as a guide to attaining therapeutic concentrations. Fluorometric liquid chromatography offers a rapid and specific method which is interference free (Figure 1).

## II. PRINCIPLE

Propranolol and added internal standard are extracted from alkalinized serum with methylene chloride and back extracted into phosphoric acid. The phosphoric acid, containing the drugs, is injected onto a reversed-phase column and eluted with a acetonitrile-tethydrofuran mobile phase. The drugs are detected fluorometrically and quantitated by peak height ratios.

## III. MATERIALS AND METHODS

### A. Equipment
A liquid chromatography (LC) system equivalent to the following is recommended. A Series I pump, an LC-100 oven, a 650-10 LC fluorescence detector (Perkin-Elmer Corp., Norwalk, Conn.), a Rheodyne® 7105 injection valve (Rheodyne, Cotati, Calif.), an Altex Ultrasphere® Octyl column, 25 cm length × 4.6 mm (Altex Scientific, Berkeley, Calif.), and a 10 mV recorder are used.

### B. Reagents
Hydrochloric acid (0.1 mol/$\ell$) reagent grade. Sodium hydroxide (1 mol/$\ell$) reagent grade. Phosphoric acid (0.043 $N$) reagent grade (equivalent to 0.1%). Phosphate buffer (1.0 mol/$\ell$ pH 4.4). Mobile phase: combine 300 m$\ell$ of acetonitrile, 50 m$\ell$ of tetranhydrofuran, and 640 m$\ell$ of water. Add 10 m$\ell$ of 1 mol/$\ell$ phosphate buffer, pH 4.4, mix and adjust PH down to 3.5 with concentrated phosphoric acid. Acetonitrile and tetrahydrofuran were UV grade from Burdick and Jackson Laboratories Inc., Muskegon, Mich.

### C. Standards
Prepare a 100 µg/m$\ell$ solution of propranolol in methanol. The propranolol was obtained from Ayerst Laboratories, Inc., New York. A working propranolol standard in serum (100 ng/m$\ell$) is prepared by mixing an appropriate amount of the stock propranolol with drug-free serum. The working standard is stable for 2 weeks at 4°C, the stock standard is stable for 1 year at 4°C. Prepare a 100 µg/m$\ell$ solution of pronetholol, the internal standard, in 0.1 mol/$\ell$ hydrochloric acid. The pronetholol was obtained from Ayerst Laboratories. A working internal standard (250 ng/m$\ell$) is prepared in water. Both the stock and working internal standard solutions are stable for 1 year at 4°C.

### D. Procedure
Pipette 200 µ$\ell$ each of the working standard and the unknowns into separate centrifuge tubes. Add 200 µ$\ell$ of the working internal standard to each tube, following by 1 drop of 1 mol/$\ell$ sodium hydroxide. Add 1 m$\ell$ of methylene chloride to each tube and vortex for 2 min. Centrifuge briefly to separate the phases and aspirate the upper aqueous layer to waste. Decant the remaining methylene chloride into a clean tube, add 100 µ$\ell$ of 0.043 $N$ phosphoric acid and vortex for 2 to 3 min. After phase separation (3 to 5 min) inject 20 µ$\ell$ of the upper phosphoric acid layer onto the column. Elute at a flow rate of 3 m$\ell$/min at a column

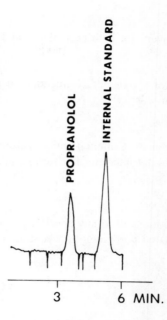

FIGURE 1. Serum containing 60 ng of propranolol per milliliter (obtained in the authors' laboratory).

temperature of 50°C. The excitation wavelength is 290 nm and the emission wavelength is 350 nm.

### E. Calculations

Calculate response factor (RF) from the working standard chromatogram as follows:

$$\frac{\text{peak height of Int. Std.}}{\text{peak height of propranolol}} = \text{Response factor (RF)}$$

Calculate each unknown from the response factor and their respective unknown chromatogram as follows:

$$\frac{\text{peak height of unknown propranolol}}{\text{peak height of Int. Std.}} \times \text{RF} \times 100 = \text{propranol concentration in ng/m}\ell$$

## IV. RESULTS

### A. Optimization of Chromatography

The pH of the mobile phase is critical in preventing the interference of dihydroquinidine with the internal standard pronetholol. If the presence of dihydroquinidine is not expected in either patient or control samples, a stronger mobile phase such as 35% acetonitrile might be used to shorten analysis time. The tetrahydrofuran eliminates peak tailing and also improves the resolution of dihydroquinidine from pronetholol.

### B. Linearity

The method is linear to at least 500 ng/m$\ell$.

## C. Recovery

Absolute and analytical recovery for both propranolol and pronetholol exceed 95%.

## D. Reproducibility

Within-run precision (CV) for a series of 20 injections was 4.2%. Day-to-day precision (CV) over a period of one year was 9.0%.

## E. Interference

There are no known interferences. Common anticonvulsants, benzodiazepines, other antiarrhythmics, and tricyclics did not interfere. Dihydroquinidine may interfere if the mobile phase is not adjusted to pH 3.5.

## F. Accuracy

This method has not been compared to any specific preexisting method, but results are in good agreement with the stated propranolol concentrations in two different commercial assayed controls.

# REFERENCE

1. **Ambler, P. K., Singh, B. N., and Lever, M.**, A simple and rapid fluorometric method for the estimation of propranolol in blood, *Clin. Chim. Acta,* 54, 373, 1974.

Chapter 11

## QUINIDINE BY FLUORESCENCE DETECTION

**George R. Gotelli and Jeffrey H. Wall**

## TABLE OF CONTENTS

| | | |
|---|---|---|
| I. | Introduction | 58 |
| II. | Principle | 58 |
| III. | Materials and Methods | 58 |
| | A. Equipment | 58 |
| | B. Reagents | 58 |
| | C. Standards | 58 |
| | D. Procedure | 58 |
| | E. Calculations | 59 |
| IV. | Results | 60 |
| | A. Optimization of Chromatography | 60 |
| | B. Linearity | 60 |
| | C. Recovery | 60 |
| | D. Reproducibility | 60 |
| | E. Interferences | 60 |
| | F. Accuracy | 60 |
| V. | Comments | 60 |
| References | | 61 |

## I. INTRODUCTION

Quinidine, the stereoisomer of quinine, has been used extensively as an antiarrhythmic agent since about 1918. Because of its toxicity and narrow therapeutic index, the measurement of the serum levels is an important part of establishing a therapeutic dosing regimen.

Five metabolites of quinidine have been identified in serum.[1] The major metabolite, 3-OH-quinidine, has antiarrhymic activity in animals and presumably also in man.[2] Commercial preparations of quinidine may contain up to 20% dihydroquinidine as a contaminant. Dihydroquinidine also has antiarrhythmic activity in man. Older quinidine assays which measure total fluorescence over-estimate the quinidine level by comeasuring these metabolities with quinidine.[3]

A more specific method which will separate quinidine from its major metabolites and contaminants is desirable (Figure 1). Liquid chromatography (LC) offers this specificity. The described method is a modification of the method of Kabra et al.[4]

## II. PRINCIPLE

Quinidine and added internal standard are extracted from alkalinized serum into methylene chloride and isopropanol. The drugs are back-extracted into phosphoric acid and an aliquot of the phosphoric acid is injected onto a reversed-phase octyl column and eluted with an acetonitrile:tetrohydrofuran:phosphate buffer mobile phase. The eluted drugs are detected by fluorometry and quantitated by peak height ratios (Figure 1).

## III. MATERIALS AND METHODS

### A. Equipment

A liquid chromatography system equivalent to the following is recommended. A Series 1 pump, LC-100 oven, 650-10LC fluorescence detector, a model 123 10MV strip-chart recorder (Perkin-Elmer Corp., Norwalk, Conn.), a Rheodyne® 7105 injection valve (Rheodyne, Cotati, Calif.), and an Altex Ultrasphere® Octyl column, 25 cm length by 4.6 mm i.d. (Altex Scientific, Berkeley, Calif.).

### B. Reagents

Phosphoric acid (0.043 $N$) reagent grade, equivalent to 0.1% solution. Phosphate buffer (1 mol/$\ell$, pH 4.4). Dissolve 136 g of potassium dihydrogen phosphate in 1$\ell$ of distilled water. Isopropanol/methylene chloride, 1:9 by volume. Sodium hydroxide, a 1 mol/$\ell$ aqueous solution. Mobile phase: 30% acetonitrile and 5% tetrahydrofuran in 10 mmol/$\ell$ phosphate buffer; combine 300 m$\ell$ of acetonitrile, 50 m$\ell$ tetrahydrofuran, 10 m$\ell$ of 1 mol/$\ell$ phosphate buffer, pH 4.4, and 630 m$\ell$ of water. Mix well. Adjust the final pH to 3.5 with phosphoric acid. The acetonitrile and tetrahydrofuran are UV grade from Burdick and Jackson Laboratories Inc., Muskegon, Mich.

### C. Standards

Quinidine standard (stock) 1000 μg/m$\ell$ in water. Pronetholol (stock internal standard) 100 μg/m$\ell$ pronetholol in 0.1 $N$ hydrochloric acid. Quinidine working serum standard. Prepare a 5 μg/m$\ell$ standard in drug-free serum from the stock standard. This serum standard is stable for 1 week at 4°C. Working internal standard. Prepare 2.5 μg/m$\ell$ pronetholol in water from the stock standard. Stable 6 months at 4°C.

### D. Procedure

Pipette 100 μ$\ell$ each of working standard and unknown into separate tubes. Add 100 μ$\ell$

FIGURE 1. Serum containing 1.7 µg/mℓ quinidine (obtained in the authors' laboratory).

of working internal standard to each tube. Add 1 drop of 1 mol/ℓ sodium hydroxide and 1 mℓ of 10% isopropanol/methylene chloride to each tube. Vortex or shake for 2 to 3 min. Centrifuge to separate the layers and aspirate to waste the upper aqueous layer. Decant the remaining organic layer into clean tubes and add 200 µℓ of 0.1% phosphoric acid to each. Vortex or shake 2 to 3 min, centrifuge briefly to separate layers, and inject 20 µℓ of the upper phosphoric acid layer onto the column. HPLC conditions are oven temperature 50°C, flow rate 3.0 mℓ/min, fluorometer exciation 290 nm, fluorometer emission 350 nm.

### E. Calculations

Calculate a response factor (R.F.) from the standard chromatogram.

$$\frac{\text{peak height of Int. Std.}}{\text{peak height of quinidine std.}} = \text{R.F.}$$

Calculate each unknown from the respective chromatogram and the R.F.

$$\frac{\text{peak height of unknown quinidine}}{\text{peak height of Int. Std.}} \times \text{RF} \times 5 = \frac{\text{quinidine conc.}}{\text{in } \mu\text{g/m}\ell}$$

## IV. RESULTS

### A. Optimization of Chromatography

The use of tetrahydrofuran in the mobile phase improves resolution and eliminates tailing without the use of high ionic strength buffers. The pH of 3.5 is critical for the separation of the contaminant dihydroxyquinidine from the pronetholol internal standard.

### B. Linearity

The method is linear to at least 50 $\mu$g/m$\ell$.

### C. Recovery

Both absolute and analytical recovery of quinidine and pronetholol exceed 95%.

### D. Reproducibility

Within run reproducibility (CV) was $\pm 3.4\%$. Day-to-day reproducibility (CV) over a period of more than 1 year was $\pm 4.5\%$.

### E. Interferences

The $N$-oxide metabolite of quinidine will coelute with quinidine with this mobile phase. This is a relatively minor metabolite which also has antiarrhythmic properties. Other common antiarrhythmics, anticonvulsants, analgesics, tricyclics, and benzodiazepines do not interfere.

### F. Accuracy

This method has not been directly compared with any preexisting methods. However, a regression analysis comparing our results with the target values on ten College of American Pathologists (CAP) and Center for Disease Control (CDC) quality control surveys gives these results: r = 0.982, slope = 1.135, and y-intercept = 0.497.

## V. COMMENTS

This method is capable of resolving 3-hydroxy quinidine and dihydroxquinidine from the parent drug quinidine. The $N$-oxide metabolite elutes simultaneously with quinidine. Back-extraction of quinidine into phosphoric acid minimizes the adsorption problems often encountered when organic solutions of basic drugs are evaporated to dryness.

## REFERENCES

1. **Rahkit, A., Kunitani, M., Holford, N. H. G., and Riegelman, S.,** Improved liquid chromatographic assay of quinidine and its metabolites in biological fluids, *Clin. Chem.*, 28, 1505, 1982.
2. **Conn, H. L. and Luichi, R. J.,** Some cellular and metabolic considerations relating to the action of quinidine as a prototype antiarrhythmic agent, *Am. J. Med.*, 37, 685, 1964.
3. **Huffman, D. H. and Hignite, C. E.,** Serum quinidine concentrations: comparison of fluorescence, gas chromatography and gas chromatographic/mass spectrometric methods, *Clin. Chem.*, 22, 810, 1976.
4. **Kabra, P. M., Chen, S. H., and Marton, L. J.,** Liquid-chromatographic determination of antiarrhythmic drugs; procainamide, lidocaine, quinidine, disopyramide and propranolol, *Ther. Drug Monit.*, 3, 91, 1981.

Chapter 12

# QUINIDINE AND ITS MAJOR METABOLITES BY FLUORESCENCE DETECTION

**Theodor W. Guentert**

## TABLE OF CONTENTS

I. Introduction .................................................................. 64

II. Principle of Method ......................................................... 64

III. Instrumentation, Materials and Methods ................................. 64
    A. Equipment ............................................................. 64
    B. Eluent .................................................................. 64
    C. Reagents and Solutions ............................................. 64
    D. Standards .............................................................. 65
    E. Procedure .............................................................. 66
    F. Calculations ........................................................... 66

IV. Results ........................................................................ 66
    A. Linearity ............................................................... 66
    B. Sensitivity ............................................................. 66
    C. Interference ........................................................... 67
    D. Precision and Bias .................................................... 67

V. Comments .................................................................... 67

References ......................................................................... 69

## I. INTRODUCTION

Quinidine is still widely used as an antiarrhythmic drug. It is extensively metabolized in man and animals. The major metabolites found in plasma have been identified as 3-OH-quinidine,[1,2] quinidine-N-oxide,[3] and 2'-quinidinone[4] (for structures see Figure 1). Furthermore, it is also likely that quinidine-10,11-dihydrodiol, a metabolite identified in rat urine,[5] occurs in humans. O-Desmethylquinidine has also been identified as a trace metabolite exclusively in urine.[6] Evidence for antiarrhythmic activity of several metabolites was elucidated from animal models,[7] and a comparison of the effects of quinidine after i.v. and oral doses to man suggested that the clinical effect is mediated not only by parent compound but also by metabolites.[8] However, the relative contribution of each metabolite to an observed effect remains to be established. Simultaneous but separate quantitation of quinidine and its metabolites 3-OH-quinidine, quinidine-N-oxide, 2'-quinidinone can be achieved by isocratic reverse-phase high-pressure liquid chromatography on an alkyl-phenyl column.[9]

## II. PRINCIPLE OF METHOD

Plasma samples are mixed with an aqueous solution containing pronethalol as an internal standard and are then buffered to pH 9. Quinidine and its metabolites are extracted by a dichloromethane-isopropanol mixture and the organic extract is evaporated to dryness. The residue is reconstituted in eluent and an aliquot is injected on a µBondapak® alkyl-phenyl column. The column effluent is monitored by fluorescence detection. Quinidine and its metabolites are quantitated by comparison of the detector response (peak height or peak area) ratio of drug (metabolite)/internal standard in the samples to that in identically prepared plasma standards containing known amounts of the compounds to be quantitated.

## III. INSTRUMENTATION, MATERIALS AND METHODS

### A. Equipment

For sample work-up, a vortexing mixer (e.g., Vortex-Genie®, Scientific Industries Inc., Bohemia, N.Y.) and a nitrogen evaporator (e.g., N-Evap®, Organomation, Northborough, Mass.) is required. For chromatography, equipment equivalent to the following should be used: Varian® Model 8500 (Varian, Palo Alto, Calif.), LDC Constametric® III (Milton Roy Co., Riviera Beach, Fla.) high-performance solvent delivery pump; Rheodyne® 71 injection valve (Rheodyne, Cotati, Calif.); alkyl-phenyl µBondapak® column (particle size, 10 µm) of 30 cm length and 3.9 mm i.d. (Waters Associates, Inc., Milford, Mass.) with an efficiency of ≥3000 plates (test compound:acenaphtene, determination by 5 sigma method); Schoeffel® FS 970 LC fluorescence detector (Schoeffel Instrument Corp., Westwood, N.J.), excitation at 245 nm, emission at 340 nm (cutoff-filter); Linear® double pen recorder (Tegal Scientific Inc., Martinez, Calif.) or Spectra-Physics® computing integrator model SP 4100 (Spectra Physics, Santa Clara, Calif.).

### B. Eluent

A mixture of 0.05 $M$ phosphate buffer pH 4.75-acetonitrile-tetrahydrofuran (80:15:5; v/v/v) is used as the eluent at a flow rate of 1.5 mℓ/min.

### C. Reagents and Solutions

All solvents are of analytical grade (except acetonitrile, tetrahydrofuran:UV grade) and were supplied by Burdick and Jackson Labs (Muskegon, Mich.), except for dichloromethane and isopropanol, which were obtained from Mallinckrodt (St. Louis).

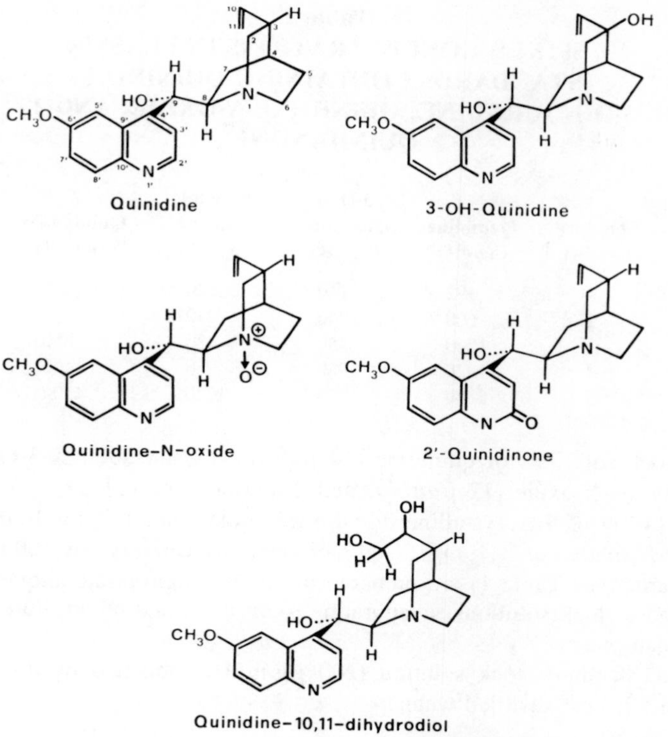

FIGURE 1. Structure of quinidine and quinidine metabolites.

1. Dichloromethane/isopropanol, 80:20% by volume.
2. Borate buffer (approximately 0.6 $M$, pH 9.0): 37.09 g $B(OH)_3$ is dissolved in 1.0 $\ell$ distilled water. The pH is adjusted to 9.0 by addition of 1 $M$ sodium hydroxide solution.
3. Phosphate buffer (0.05 $M$, pH 4.75): 6.80 g $KH_2PO_4$ is dissolved in 1.0 $\ell$ of distilled water and filtered through a 0.45 μm Millipore® filter type HA (Millipore Corp., Bedford, Mass.).
4. Mobile phase: to prepare 1 $\ell$ of the HPLC mobile phase 800 m$\ell$ 0.05 $M$ phosphate buffer pH 4.75 is mixed with 150 m$\ell$ acetonitrile. This binary mixture is degassed by evacuation for 1 min following which 50 m$\ell$ tetrahydrofuran is added to obtain the ternary mixture. The solvent is degassed once more by flash evacuation followed by ultrasonication and allowed to stand overnight before use.

## D. Standards

Commercially available quinidine (e.g., J. T. Baker, Phillipsburg, N.J.) contains approximately 4 to 8% of dihydroquinidine as an impurity; up to 20% of the dihydro product is acceptable in pharmaceutical preparations according to pharmacopeial specifications.[10] A dihydro-free quinidine standard was prepared by removing dihydroquinidine according to the method described by Thron and Dirscherl.[11] Crystallization from anhydrous ethanol yielded crystalline quinidine with 1 mol of ethanol. The 3-OH-quinidine metabolite and quinidine-N-oxide were synthesized from quinidine according to published procedures.[3,12] The metabolite 2'-quinidinone was a gift from Triangle Research Institute (Research Triangle Park, N.C.) and the internal standard used, pronethalol, was kindly supplied by ICI® (Macclesfield, Cheshire, U.K.)

## Table 1
## SPIKED CONCENTRATIONS IN PLASMA STANDARDS CONTAINING QUINIDINE, 3-OH-QUINIDINE, QUINIDINE-$N$-OXIDE, AND 2'-QUINIDINONE

| Standard number | Quinidine (ng/m$\ell$) | 3-OH-Quinidine (ng/m$\ell$) | Quinidine-$N$-oxide (ng/m$\ell$) | 2'-Quinidinone (ng/m$\ell$) |
|---|---|---|---|---|
| 1 | 192 | 120 | 120 | 110 |
| 2 | 600 | 240 | 240 | 275 |
| 3 | 1200 | 480 | 480 | 550 |
| 4 | 1920 | 960 | 960 | 1100 |
| 5 | 3840 | 1920 | 1920 | 1100 |

Separate stock solutions of quinidine (24 µg/m$\ell$), the metabolites 3-OH-quinidine (12 µg/m$\ell$), quinidine-$N$-oxide (12 µg/m$\ell$), and 2'-quinidinone (11 µg/m$\ell$) are prepared by dissolving 2.74 mg of the crystalline dihydro-free quinidine, 1.2 mg 3-OH-quinidine, 1.2 mg quinidine-$N$-oxide, or 1.1 mg 2'-quinidinone, respectively, in 100.0 m$\ell$ methanol. Plasma standards (see Table 1) are prepared by mixing appropriate aliquots of methanolic dilutions of these stock solutions, evaporating to dryness, and reconstituting the residue in drug-free human plasma.

The internal standard stock solution (200 µg/m$\ell$) is obtained by dissolving 10.0 mg pronethalol in 50.0 m$\ell$ distilled water.

### E. Procedure

Mix 200 µ$\ell$ of plasma sample or plasma standard with 200 µ$\ell$ aqueous pronethalol solution 200 µg/m$\ell$ and 200 µ$\ell$ of 0.6 $M$ borate buffer pH 9.0. Extract with 10 m$\ell$ dichloromethane-isopropanol 4:1 (v/v) by vortexing for 1 min. Centrifuge the mixture for 5 min at 600 to 1200 g and transfer the organic layer to a separate test tube. Evaporate to dryness, reconstitute the residue in 200 µ$\ell$ of eluent, and inject a 50 µ$\ell$ aliquot onto the HPLC column. Detector sensitivity is set to 0.05 µA full scale and the time constant at 4 sec. For low plasma levels of quinidine and metabolites, the volume of plasma taken for assay and of buffer added can be increased from 200 µ$\ell$ to 1.0 m$\ell$.

### F. Calculations

Plasma standards (Table 1) are carried through the procedure and a least squares unweighted regression line of peak or area ratios drug/internal standard is calculated separately for quinidine and each of the metabolites. Unknown concentrations in specimens are calculated from the parameters defining the regression lines.

## IV. RESULTS

### A. Linearity

Peak height ratios of drug to internal standard are linearly related in the therapeutically relevant concentration range covered by the standards in Table 1. The coefficients of variation for concentration-normalized peak height ratios were found to be <3% for all compounds except for 2'-quinidinone (<8%).

### B. Sensitivity

The detection limit for all compounds in the assay is approximately 10 to 20 ng/m$\ell$ when extracting volumes of 0.5 to 1 m$\ell$ of specimen.

## Table 2
## INTERFERENCE BY COMMONLY USED DRUGS

| Drug | Conc. spiked (mg/ℓ) | 2'-Quinidinone | 3-OH-Quinidine | Quinidine | Quinidine-N-oxide |
|---|---|---|---|---|---|
| Salicylic acid | 1200 | xxxx | xxxx | | |
| Alprenolol | 150 | | | | xx |
| Propranolol | 0.3 | | | | xx |
| Phenprocoumon | 60 | | | | xx |
| Imipramine | 0.45 | | | | xx |
| Diazepam | 9 | | x | | |
| Prazosin | 90 | xxxx | xxxx | xxxx | xxxx |
| Theophylline | 60 | | xx | | |

*Note:* Quantitative assessment of interference: interference corresponds to the following concentrations of quinidine or metabolite.

| | | |
|---|---|---|
| 2'-Quinidinone | xxxx | > 200 ng/mℓ |
| 3-OH-quinidine/ | x | < 30 ng/mℓ |
| quinidine-N-oxide | xx | 30—150 ng/mℓ |
| | xxxx | > 300 ng/mℓ |
| Quinidine | xxxx | > 1000 ng/mℓ |

### C. Interference

A series of commonly used drugs were tested for interference with the assay at concentrations three times the upper limit of their accepted therapeutic range. The following compounds tested were found not to present any problems in the described quantitation of quinidine and metabolites — analgesics: phenacetin, indomethacin, morphine; anticoagulant: warfarin; psychotropics: desipramine, chlordiazepoxide; diuretics:spironolactone, furosemide; cardioactive drugs: verapamil, procainamide, mexiletin, disopyramide, lidocaine, dihydralazine, digoxin, nifedipin; hypolipemic drug: clofibrate; $H_2$-blocker: cimetidine. The most severe interference occurred with high concentrations of salicylic acid and prazosin. Table 2 lists the drugs found to interfere with the assay and gives indications on the quantitative relevance of this interference.

### D. Precision and Bias

The within-day and day-to-day precision for quinidine and the metabolites in the assay is better than a 6% coefficient of variation. Bias is minimal; in spiked samples assayed concentrations of all compounds were no more than 2% off the expected values.

## V. COMMENTS

The HPLC assay procedure described allows separation of quinidine from its dihydro product and from its major metabolites using isocratic conditions (Figure 2). The separation of the complex mixture requires careful optimization of the chromatographic conditions and the assay should therefore be reserved for instances where quantitation of the active metabolites seems important; for instance, in studies concerned with correlations of pharmacodynamic effects with plasma concentrations, in patients where altered metabolism is suspected, or when an unexpected response to concentrations of unchanged drug is observed.

The pH value in the eluent was found to be a crucial factor for successful separation of

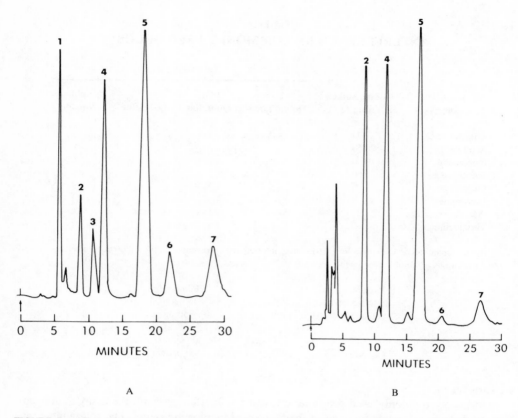

FIGURE 2. Chromatograms of (A) plasma spiked with quinidine and metabolites and (B) of plasma from a patient who received multiple doses of quinidine. Key: (1) 2'-quinidinone, (2) 3-OH-quinidine, (3)$O$-desmethylquinidine, (4) pronethalol (internal standard), (5) quinidine, (6) dihydroquinidine, (7) quinidine-$N$-oxide.

all the substances measured. At low pH values (pH 2.5 to 3.5) quinidine-$N$-oxide was not separable from quinidine using various buffer systems and organic modifiers. In the higher pH range (pH 5 to 7) the two compounds eluted with different retention times but base-line separation of 2'-quinidinone and 3-OH-quinidine was not possible. The intermediate pH-value employed represents a compromise to solving the two separation problems. Tetrahydrofuran is added in small amounts to the solvent because it eliminates tailing and promotes sharper peaks, resulting in a better separation. Depending on the column performance, small changes in acetonitrile and tetrahydrofuran content in the eluent may be necessary to obtain satisfactory separation of all compounds. Slightly elevated column temperatures (30 to 40°C) are another means of achieving the same goal.

The commonly accepted therapeutic concentration range of quinidine in plasma of 2 to 6 µg/mℓ was based on studies by Sokolow and Edgar.[13] However, these authors employed a nonspecific fluorescence procedure. Steady-state concentrations of quinidine determined using specific assay procedures are much lower (1 to 3 µg/mℓ) after therapeutically effective quinidine doses.

# REFERENCES

1. **Carroll, F. I., Smith, D., and Wall, M. E.**, Carbon-13 magnetic resonance study. Structure of the metabolites of orally administered quinidine in humans, *J. Med. Chem.*, 17, 985, 1974.
2. **Beermann, B., Leander, K., and Lindström, B.**, The metabolism of quinidine in man: structure of a main metabolite, *Acta Chem. Scand.*, 30, 465, 1976.
3. **Guentert, T. W., Daly, J. J., and Riegelman, S.**, Isolation, characterisation and synthesis of a new quinidine metabolite, *Eur. J. Drug Metab. Pharmacokin.*, 7, 31, 1982.
4. **Brodie, B. B., Baer, J. E., and Craig, L. C.**, Metabolic products of the cinchona alcaloids in human urine, *J. Biol. Chem.*, 188, 567, 1951.
5. **Barrow, S. E., Taylor, A. A., Horning, E. C., and Horning, M. G.**, High-performance liquid chromatographic separation and isolation of quinidine and quinine metabolites in rat urine, *J. Chromatogr.*, 181, 219, 1980.
6. **Drayer, D. E., Lowenthal, D. T., Restivo, K. M., Schwartz, A., Cook, C. E., and Reidenberg, M. M.**, Steady-state serum levels of quinidine and active metabolites in cardiac patients with varying degrees of renal function, *Clin. Pharmacol. Ther.*, 24, 31, 1978.
7. **Drayer, D. E., Cook, C. E., and Reidenberg, M. M.**, Active quinidine metabolites, *Clin. Res.*, 24, 623A, 1976.
8. **Holford, N. H. G., Coates, P. E., Guentert, T. W., Riegelman, S., and Sheiner, L. B.**, The effect of quinidine and its metabolites on the electrocardiogram and systolic time intervals: concentration-effect relationships, *Br. J. Clin. Pharmacol.*, 11, 187, 1981.
9. **Guentert, T. W., Rakhit, A., Upton, R. A., and Riegelman, S.**, An integrated approach to measurements of quinidine and metabolites in biological fluids, *J. Chromatogr.*, 183, 514, 1980.
10. The United States Pharmacopeia, United States Pharmacopeial Convention Inc; Rockville, Md., 20852, 20th rev., 1979.
11. **Thron, H. and Dirscherl, W.**, Eine einfache Methode zur Trennung der China-Alkaloide von ihren Dihydrobasen, *Justus Liebigs Ann. Chem.*, 515, 252, 1935.
12. **Carroll, F. I., Philip, A., and Coleman, M. C.**, Synthesis and stereochemistry of a metabolite resulting from the biotransformation of quinidine in man, *Tetrahedron Lett.*, 21, 1757, 1976.
13. **Sokolow, M. and Edgar, A. L.**, Blood quinidine as a guide in the treatment of cardiac arrhythmias, *Circulation*, 1, 576, 1950.

Chapter 13

# SIMULTANEOUS ANALYSIS OF CARBAMAZEPINE, ETHOSUXIMIDE, PHENOBARBITAL, PHENYTOIN, AND PRIMIDONE

## Pokar M. Kabra

## TABLE OF CONTENTS

I. Introduction ................................................................. 72

II. Principle .................................................................... 72

III. Materials and Methods ..................................................... 72
    A. Equipment ............................................................ 72
    B. Reagents .............................................................. 72
    C. Standards ............................................................. 72
    D. Procedure ............................................................. 73
    E. Calculation ........................................................... 73

IV. Results ...................................................................... 73
    A. Linearity .............................................................. 73
    B. Recovery ............................................................. 73
    C. Interference .......................................................... 73
    D. Precision ............................................................. 74
    E. Temperature ......................................................... 74
    F. Metabolites .......................................................... 74

    Comments ................................................................. 74

References ....................................................................... 76

## I. INTRODUCTION

Therapeutic monitoring of anticonvulsant drugs is rapidly becoming a routine aid in the clinical management of patients with seizure disorders. While phenobarbital and phenytoin remain the two most frequently prescribed anticonvulsant drugs, patients frequently receive others. Thus, comprehensive monitoring of anticonvulsant therapy demands the availability of suitable methods for separating and determining various combinations of these drugs and their bioactive metabolites. In general, gas-liquid chromatography and liquid chromatography (LC) provide this capability. Gas-liquid chromatography requires a relatively large sample and significant time for sample extraction and derivatization. In contrast, the LC method described can be performed on 25 $\mu\ell$ of plasma and requires minimal sample preparation.

## II. PRINCIPLE

Serum proteins are precipitated from 25 to 500 $\mu\ell$ of serum with acetonitrile containing hexobarbital as an internal standard. After centrifugation, an aliquot of the supernatant is injected onto a reverse-phase column and the drugs are eluted with an acetonitrile-phosphate buffer-mobile phase. The drugs are detected by their absorption at 195 nm and quantified from either their peak heights or peak areas.

## III. MATERIALS AND METHODS

### A. Equipment

Liquid chromatograph systems equivalent to the following are recommended: a Model 601 (Perkin-Elmer Corp., Norwalk, Conn.) equipped with a Rheodyne® 7105 valve (Rheodyne, Berkeley, Calif.) a variable wavelength detector (Perkin-Elmer LC55 or LC65T), a temperature controlled oven (Perkin-Elmer Model LC100 or LC65T), and any one of the following reverse-phase octadecylsilane columns: Waters® C18 µBondapak® 30 cm × 4 mm i.d. (Waters Associates, Inc., Milford, Mass.), Whatman® ODS 3, 25 cm × 4.6 mm i.d. (Whatman, Inc., Clifton, N. J.), or Altex® spherisorb ODS 25cm × 4.6 mm i. d. (Altex Scientific, Berkeley, Calif.). A stainless steel precolumn (5 cm × 2.1 mm i.d. with Swagelok® fittings (Alltech Associates, Arlington Heights, Ill.) packed with 25 to 40 µm ODS (MC/B Manufacturing, Cincinnati, Ohio) is mounted in line between the column and injector. A strip-chart recorder or a digital data system (e.g., Sigma® 10 Perkin-Elmer Corp.) and a model 5412 Eppendorf® centrifuge with 1.5 m$\ell$ polypropylene tubes (Brinkman Instruments, Inc., Westbury, N. Y.) are utilized.

### B. Reagents

Acetonitrile: acetonitrile (UV grade), distilled in glass (Burdick and Jackson Laboratories, Inc., Muskegon, Mich.). Mobile phase: a solution of 95 m$\ell$ acetonitrile in 405 m$\ell$ phosphate buffer pH 4.4). Phosphate buffers: add 300 $\mu\ell$ 1 mol/$\ell$ $KH_2PO_4$ to 1800 m$\ell$ distilled water and adjust the pH to 4.4 with 0.9 mol/$\ell$ phosphoric acid. Filter through a 0.45 µm Millipore® filter (Millipore Corp., Bedford Mass.) and degas.

### C. Standards

Phenytoin (diphenylhydantoin) can be obtained from Eastman Kodak Co., Rochester, N. Y., phenobarbital from Winthrop Laboratories, New York, ethosuximide from Parke, Davis and Co., Detroit, Mich., carbamazepine from Ciba-Geigy Corp., Summit, N.J., hexobarbital from Sigma Chemical Co., St. Louis, and primidone was a gift from Ayerst Laboratories, Inc., New York.

The anticonvulsant stock standard is prepared as follows: 100 mg each of phenobarbital, phenytoin, primidone, ethosuximide, and carbamazepine (1 g/ℓ) are dissolved in 100 mℓ methanol. The solution is stable at 4°C for at least 6 months.

Anticonvulsant serum or reference standards: dilute 0.5, 1, 2, and 10 mℓ of the anticonvulsant stock solution to 100 mℓ with pooled drug-free serum or methanol to obtain concentrations of 5, 10, 20, 50, and 100 µg/mℓ, respectively.

A stock internal standard is made by dissolving 50 mg hexobarbital in 100 mℓ acetonitrile. A tenfold dilution of the stock with acetonitrile is used as the working internal standard.

Quality control specimens (Thera-Chem® tri-level anticonvulsant) can be obtained from Fisher Scientific Co., Diagnostics Division, Orangeburg, N.Y., or prepared in-house in the same manner as the serum standards.

### D. Procedure*

To 200 µℓ of serum sample, serum standard, or control in a 0.5 mℓ polypropylene tube, add 200 µℓ acetonitrile containing hexobarbital as an internal standard (sample volume can be varied between 25 and 500 µℓ as long as an equal volume of acetonitrile is added). Vortex-mix the mixture 10 sec and then centrifuge for 1 min at 10,000 × g in an Eppendorf® centrifuge. Inject approximately 20 µℓ of the supernatant onto the chromatograph and elute with the mobile phase at a flow rate of 3.0 mℓ/min. The column temperature is maintained at 50°C and the column effluent is monitored at 195 nm. Detector sensitivity is set to 0.04 A full scale. Total chromatographic time is less than 15 min (Figure 1).

### E. Calculation

1. Aqueous or serum standards are carried through the procedure and used to prepare a standard curve based on relative peak height or peak area.
2. Calculate the concentration of unknown anticonvulsant in the specimen by direct comparison with the data obtained from the standard curve of the reference or serum standards.

## IV. RESULTS

### A. Linearity

Peak height ratios of drug to to internal standard are linearly related from 2 to at least 100 µg/mℓ of each drug.

### B. Recovery

Absolute recoveries ranged from 95 to 106% and relative recoveries from 92 to 103%[1] in concentration range of 2 to 100 µg/mℓ.

### C. Interference

Of more than 30 drugs tested, only ethotoin interfered with the analysis of phenobarbital. Grossly hemolyzed, icteric, or lipemic samples can be assayed without interference.[1]

---

* The optimized chromatographic conditions for the separation of these drugs were established by careful investigation of all the analytical variables involved. The pH of the mobile phase was varied between 3 and 7; pH 4.4 was found to be optimal for the resolution of these drugs without significant interference from endogenous or exogenous compounds. Serum based standards are not necessary for quantitation in this method since solvent extraction is not employed in the assay and since no significant difference in quantifying unknowns using either aqueous standards or serum based standards was found.

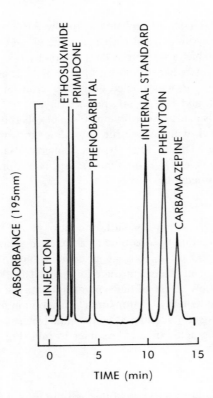

FIGURE 1. Chromatogram of a standard mixture of anticonvulsants, 75 ng of each drug other than hexobarbital (150 ng) was chromatographed.

## D. Precision

The precision for all 5 drugs equals less than 6% coefficient of variation for both within-day and day-to-day analyses.

## E. Temperature

The use of elevated temperature has a marked effect on the retention time and resolution of anticonvulsants on octadecylsilane reverse-phase columns. Elevated temperature helps reduce the viscosity of the mobile phase, resulting in a lower pressure at a given flow rate. Elevated temperature improves the efficiency and selectivity of the column. Phenytoin and carbamazepine coelute at ambient temperature but not at elevated temperature using the same chromatographic conditions. In addition to reducing analysis time, elevated temperature results in better reproducibility of retention times. Heating devices which only partially enclose the column, rather than a fully enclosed oven, are usually inadequate for the resolution of phenytoin and carbamazepine.

## F. Metabolites

Phenylethyl malonamide (a metabolite of primidone) and carbamazepine 10, 11-epoxide (a metabolite of carbamazepine) are well-separated from the other drugs and can readily be quantified if desired (Figure 2). Additionally, several other minor anticonvulsants and metabolites can be measured using this chromatographic system.[2]

## V. COMMENTS

Specimens for anticonvulsant drug analysis (preferably serum) are normally drawn prior

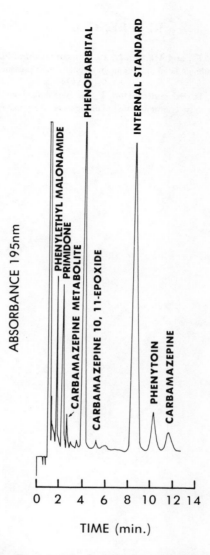

FIGURE 2. Chromatogram of a patient's serum which contained 8 μg/mℓ primidone, 36 μg/mℓ phenobarbital, 4 μg/mℓ phenytoin, and 3 μg/mℓ carbamazepine. Note the metabolites of primidone (PEMA and phenobarbital) and carbamazepine (carbamazepine 10, 11-epoxide and an unidentified metabolite).

to the next drug dose to insure that the steady state trough concentration is measured. The commonly accepted therapeutic range and serum half-life for these drugs are listed below:

| Drug | Therapeutic range (mg/ℓ) | Serum half-life (hr) |
|---|---|---|
| Carbamazepine | 4—12 | 10—30 |
| Ethosuximide | 40—100 | 40—60 |
| Phenobarbital | 15—40 | 50—120 |
| Phenytoin | 10—20 | 18—30 |
| Primidone | 5—12 | 3—12 |

## REFERENCES

1. **Kabra, P. M., Stafford, B. E., and Marton, L. J.**, Simultaneous measurement of phenobarbital, phenytoin, primidone, ethosuximide, and carbamazepine in serum by high-pressure liquid chromatography, *Clin. Chem.*, 23, 1284, 1977.
2. **Kabra, P. M., McDonald, D. N., and Marton, L. J.**, A simultaneous high-performance liquid chromatographic analysis of the most common anticonvulsants and their metabolites in the serum, *J. Anal. Toxicol.*, 2, 127, 1978.

Chapter 14

# SIMULTANEOUS VERY HIGH SPEED LIQUID-CHROMATOGRAPHIC ANALYSIS OF ETHOSUXIMIDE, PRIMIDONE, PHENOBARBITAL, PHENYTOIN, AND CARBAMAZEPINE IN SERUM

**Pokar M. Kabra**

## TABLE OF CONTENTS

| | | |
|---|---|---|
| I. | Principle | 78 |
| II. | Materials and Methods | 78 |
| | A. Equipment | 78 |
| | B. Reagents | 78 |
| | C. Standards | 78 |
| | D. Procedure 195 | 79 |
| | E. Procedure 210 | 80 |
| | F. Calculation | 80 |
| III. | Results | 80 |
| | A. Linearity | 80 |
| | B. Recovery | 80 |
| | C. Precision | 81 |
| | D. Detection | 81 |
| | E. Interference | 81 |
| | F. Temperature | 81 |
| | G. Metabolites | 81 |
| | H. Accuracy | 81 |
| IV. | Comments | 81 |
| References | | 82 |

## I. PRINCIPLE

The anticonvulsant drugs are extracted from 200 µℓ of serum containing cyclopal as an internal standard with a Bond-Elut® column. The Bond-Elut® column is eluted with 300 µℓ of methanol and an aliquot of the eluate is injected onto a reversed-phase high speed column. The anticonvulsant drugs are eluted with an acetonitrile/methanol/phosphate buffer mobile phase. The drugs are detected by their absorption at 210 or 195 nm and quantitated from either their peak height or peak area ratios.

## II. MATERIALS AND METHODS

### A. Equipment

Liquid chromatograph (LC) systems equivalent to the following are recommended. A Model Series 2 or Series 3 liquid chromatograph equipped with a Model LC-100 column oven, a Model LC-85 variable wavelength detector, and a Sigma® 10 Data system (all from Perkin-Elmer Corp., Norwalk, Conn.) can be used. A high speed chromatography package consisting of a Rheodyne® Model 7125 injector with a 6 µℓ loop, a Perkin-Elmer 2.4 µℓ micro flow cell for the LC-85 detector, and 50 cm × 0.18 mm (i.d.) stainless steel connecting tubing is used to minimize extra-column band broadening. A Perkin-Elmer column (125 × 4.6 mm, i.d.) packed with 5-µm particle size C18 reversed-phase packing or a 100 × 4.6 mm (i.d.) column packed with 3-µm particle size C18 packing (Applied Sciences Laboratories, State College, Pa.) is mounted in the oven which is maintained at 50°C. The column is eluted with methanol/acetonitrile/phosphate buffer (35/13.5/51.5, by volume) at a flow rate of 3.0 mℓ/min and the absorbance of the column effluent is monitored at 195 or 210 nm (Figures 1 and 2).

### B. Reagents

All chemicals used are of reagent grade. Acetonitrile and methanol, all distilled in glass "UV grade" are from Burdick and Jackson Laboratories, Inc., Muskegon, Mich.

Phosphate buffer, 20 mmol/ℓ, pH 3.7, is prepared by dissolving 2.7 g of $KH_2PO_4$ in 1ℓ of water. The pH is adjusted to 3.7 with phosphoric acid. Phosphate buffer, 100 mmol/ℓ, pH 4.4, is prepared by dissolving 13.6 g of $KH_2PO_4$ in 1 ℓ of water. The pH is adjusted to 4.4 with phosphoric acid. The mobile phase is prepared by mixing 350 mℓ of methanol, 135 mℓ of acetonitrile with 515 mℓ of 20 mmol/ℓ phosphate buffer. Vac-Elut® (#Al 6000) apparatus and Bond-Elut® extraction column (#607101) are obtained from Analytichem International Inc., Harbor City, Calif.

### C. Standards

Phenytoin can be obtained from Eastman Kodak Co., Rochester, N.Y.; phenobarbital from Winthrop Laboratories, New York; ethosuximide from Parke, Davis and Co., Detroit, Mich.; carbamazepine from Ciba-Geigy Corp., Summit, N.J.; primidone from Ayerst Laboratories, Inc., New York; and cyclopal from Applied Sciences Labs, Inc., State College, Pa.

The anticonvulsant stock standard is prepared by dissolving 100 mg each of ethosuximide, primidone, phenobarbital, phenytoin, and carbamazepine in 100 mℓ of methanol.[1] The solution is stable at 4°C for at least 1 year.

The stock internal standard is prepared by dissolving 50 mg of cyclopal in 100 mℓ of water. The working internal standard is prepared by diluting the stock internal standard tenfold with 100 mmol/ℓ phosphate buffer. To prepare the anticonvulsant serum or reference standards, evaporate 0.5, 1.0, 5.0, and 10.0 mℓ of the anticonvulsant stock solution and

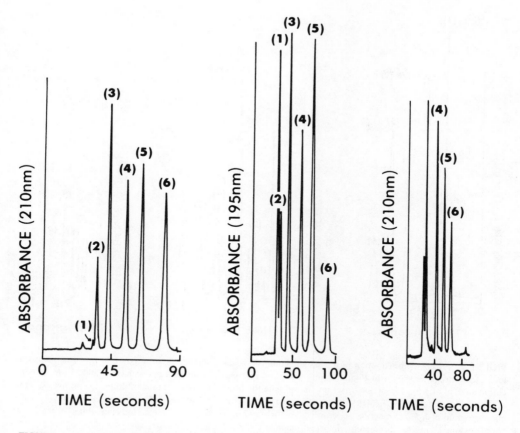

FIGURE 1. Liquid chromatograms of some antiepileptic drugs. Left: chromatogram of a 5 µℓ injection of standard anticonvulsant mixture with the following concentration of drugs (detected at 210 nm): (1) ethosuximide 20 mg/ℓ; (2) primidone 10 mg/ℓ (3) phenobarbital 20 mg/ℓ; (4) cyclopal 50 mg/ℓ (int. std.); (5) phenytoin 20 mg/ℓ; and (6) carbamazepine 10 mg/ℓ. Middle: chromatogram of a 5 µℓ of anticonvulsant standard at the same concentrations of the above drugs as shown above (detected at 195 nm). Right: chromatogram of a 5 µℓ injection of serum sample with 20 mg/ℓ of phenobarbital and 14 mg/ℓ of phenytoin. Chromatograms were obtained using a 3µm particle size reversed-phase column. (From Kabra, P. M., Nelson, M. A., and Marton, L. J., *Clin. Chem.*, 29, 473, 1983. With permission.)

reconstitute with 100 mℓ of drug-free serum to obtain drug concentrations of 5, 10, 20, 50, and 100 mg/ℓ, respectively.

## D. Procedure 195*

Place Bond-Elut® columns on the top of the Vac-Elut® chamber. Connect the Vac-Elut® chamber to vacuum and pass two column volumes of methanol and distilled water through each column. After disconnecting the vacuum, place 200 µℓ of working internal standard (cyclopal, 50 µg/mℓ) onto each column, then pipette 200 µℓ of standard, control, or patient's serum onto each labeled column. Connect the vacuum to the Vac-Elut® chamber again and wash each column with 2 vol of distilled water. Disconnect the vacuum and place a rack of labeled 10 × 75 mm glass tubes in the Vac-Elut® chamber under each corresponding Bond-Elut® column. Pipette 300 µℓ of methanol onto each column and connect the vacuum again. After the eluate is collected, remove the rack of tubes from the chamber, shake the tubes to mix the eluate, and inject 5 µℓ from each tube onto the liquid chromatograph.

* Applicable to all five drugs, absorbance measured at 195 nm.

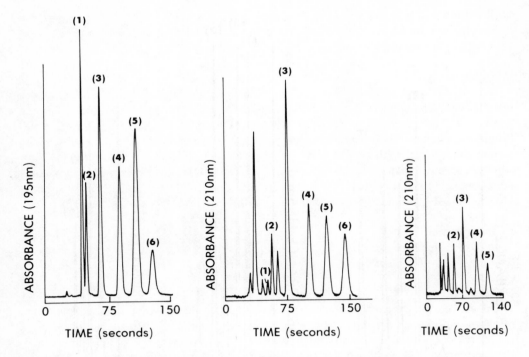

FIGURE 2. Left: chromatogram of a 5 µℓ injection of standard anticonvulsant mixture at the same concentrations of the drugs as shown in Figure 1 left. The flow rate was 3.5 mℓ/min and detection at 195 nm. Middle: chromatogram of a 5 µℓ injection of serum control with the following concentration of drugs: (2) primidone 7 mg/ℓ; (3) phenobarbital 30 mg/ℓ; (4) cyclopal 50 mg/ℓ; (5) phenytoin 15 mg/ℓ; and (6) carbamazepine 6 mg/ℓ. Right: chromatogram of a 5 µℓ injection of serum sample with: (2) primidone 11 mg/ℓ; (3) phenobarbital 20 mg/ℓ; and (5) phenytoin 11 mg/ℓ. Chromatograms were obtained using a 5 µm particle size reversed-phase column. (From Kabra, P. M., Nelson, M. A., and Marton, L. J., *Clin. Chem.*, 29, 473, 1983. With permission.)

### E. Procedure 210
Procedure 210 is identical to procedure 195 except that the detector wavelength is set at 210 nm instead of 195 nm. This procedure can be used if ethosuximide is not analyzed.

### F. Calculation

1. Serum standards are carried through the procedure and used to prepare a standard curve based on relative peak height or peak area.
2. Calculate the concentration of unknown anticonvulsant in the specimen by direct comparison with the data obtained from the standard curve of serum standards.

## III. RESULTS

### A. Linearity
Peak area ratios of drug to internal standard are linearily related from 5 to at least 200 mg/ℓ of each drug.

### B. Recovery
Analytical recoveries ranged from 92 to 109% in a concentration range of 5 to 200 mg/ℓ.

## C. Precision

The precision for all five drugs equals less than 5% coefficient of variation for both within-day and day-to-day analyses.

## D. Detection

The variable wavelength detector is set at 210 nm for the simultaneous analysis of primidone, phenobarbital, phentoin, and carbamazepine. However, because ethosuximide absorbs poorly at 210 nm, the detector wavelength is set at 195 nm for the analysis of ethosuximide. Primidone, phenobarbital, phenytoin, and carbamazepine absorb well at both wavelengths, so that either wavelength may be used for their detection. Except for carbamazepine, sensitivity is greatly enhanced for all of these drugs at 195 nm. However, because of limitations of detection at 195 nm — including decreased specificity and higher background absorption — detection at 210 nm is preferred for the routine analysis of anticonvulsant drugs.

## E. Interference

Of more than 70 drugs tested, only phensuximide, pentobarbital, and glutethimide interfere with the analysis of phenobarbital and carbamazepine.

## F. Temperature

The use of elevated temperature has a marked effect on the retention times and resolution of anticonvulsants on octadacylsilane reversed-phase columns. Elevated temperature helps reduce the viscosity of the mobile phase, resulting in a lower pressure at a given flow rate. Elevated temperature improves the efficiency and selectivity of the column. In addition to reducing analysis time, elevated temperature results in better reproducibility of retention time.

## G. Metabolites

Phenylethyl malonamide (a metabolite of primidone), 5-hydroxyphenyl-5-phenylhydantoin (a metabolite of phenytoin), and carbamazepine 10,11 epoxide (a metabolite of carbamazepine) are well-separated from the other drugs and can readily be quantitated if desired.

## H. Accuracy

The high-speed liquid chromatographic method correlated well with an established light chromatographic method.[2] The correlation coefficient ranged from 0.969 to 0.997 for the five drugs.

## IV. COMMENTS

The very high speed liquid chromatographic analysis of anticonvulsants require four specific modifications of conventional liquid chromatographic systems. A small particle packing (3 or 5 μm) results in higher efficiency than conventional packing (10 μm). These shorter columns reduce analysis time because of the smaller column void volume. Extra-column band broadening is significantly reduced by using shorter, small bore (0.18 mm) connecting tubing and UV detector with a small volume flow cell (2.4 μℓ). These modifications reduce band broadening to about one fourth of that seen with conventional liquid chromatographs. Use of a detector with a very fast response time (135 msec) allows for accurate detection of fast eluting analytes. The Bond-Elut® reversed-phase columns facilitate rapid extraction of anticonvulsant drugs from serum with very high efficiency and precision. The Bond-Elut® columns can be regenerated for repetitive use by passing two column volumes of methanol through them; 15 to 20 serum samples could be processed without appreciable loss in the extraction efficiency.

## REFERENCES

1. **Kabra, P. M., Nelson, M. A., and Marton, L. J.,** Simultaneous very fast liquid-chromatographic analysis of ethosuximide, primidone, phenobarbital, phenytoin and carbamazepine in serum, *Clin. Chem.*, 29, 473, 1983.
2. **Kabra, P. M., Stafford, B. E., and Marton, L. J.,** Simultaneous measurement of phenobarbital, phenytoin, primidone, ethosuximide, and carbamazepine in serum by high-pressure liquid chromatography, *Clin. Chem.*, 23, 1284, 1977.

Chapter 15

# SIMULTANEOUS ANALYSIS OF COMMON SEDATIVES IN SERUM

**Pokar M. Kabra**

## TABLE OF CONTENTS

I. Introduction .................................................................. 84

II. Principle ..................................................................... 84

III. Materials and Methods ........................................................ 84
    A. Equipment ............................................................... 84
    B. Reagent ................................................................. 84
    C. Standards ............................................................... 84
    D. Procedure ............................................................... 85
    E. Calculation ............................................................. 85

IV. Results ...................................................................... 85
    A. Linearity ............................................................... 85
    B. Recovery ................................................................ 85
    C. Precision ............................................................... 85
    D. Interference ............................................................ 85

V. Comments .................................................................... 87

References ....................................................................... 87

## I. INTRODUCTION

Rapid identification and quantitation of barbiturates and hypnotics may be helpful in managing comatose patients with suspected drug overdose. Various techniques currently used for screening these drugs include spectrophotometry, paper chromatography, thin-layer chromatography, gas chromatography, gas chromatography mass-spectrometry, immunoassays, and liquid chromatography.[1] Spectrophotometric analysis often is time consuming, lacks specificity and sensitivity, and is usually applicable only to measurement of a single drug. Thin layer and paper chromatography are valuable techniques for the detection of multiple drugs. However, these techniques are nonspecific, time consuming, and provides only semiquantitative analysis. Gas chromatography and gas chromatography-mass spectrometry can resolve multiple drugs; however, the sample extraction methods and derivatization procedures often employed for the analysis of these compounds are complex and time consuming. Immunological assays provide high sensitivity and speed, but with serious limitation in specificity.

Liquid chromatography (LC) offers some advantages in analyzing for these drugs in serum. Drugs may be analyzed without derivatization and the drugs may be collected unchanged in the column eluate for further characterization. A simple LC method for the simultaneous analysis of 12 hypnotic drugs is presented here.[2]

## II. PRINCIPLE

Serum proteins are precipitated from 25 to 500 $\mu\ell$ of serum with acetonitrile containing 5-(4-methylphenyl)-5-phenylhydantoin as an internal standard. After centrifugation, an aliquot of the supernatant is injected into a reversed-phase column and the drugs are eluted with an acetonitrile-phosphate buffer mobile phase. The drugs are detected by their absorption at 195 nm and quantitated from their peak height ratios.

## III. MATERIALS AND METHODS

### A. Equipment

Liquid chromatograph system equivalent to the following are recommended: Model 601, Series 2 or Series 3 liquid chromatograph equipped with a Rheodyne® 7105 valve (Rheodyne, Cotati, Calif.); a variable-wavelength detector (LC-75 or LC 65T); a temperature controlled oven (Perkin-Elmer Model LC 100 or LC 65T) are used. The prepacked reversed-phase column (µBondapak® C18, 30 cm × 4 mm, Waters Associates, Inc., Milford, Mass.) is mounted in the oven. A strip-chart recorder or a digital data system (e.g., Sigma® 10 Perkin-Elmer Corp., Norwalk, Conn.) and a Model 5412 Eppendorf® centrifuge with 1 · 5 m$\ell$ polypropylene tubes (Brinkman Instruments, Inc., Westbury, N.Y.) are used.

### B. Reagent

Acetonitrile (UV grade), distilled in glass (Burdick and Jackson Laboratories, Inc., Muskegon, Mich.); phosphate buffer (pH 4.4) is prepared by dissolving 300µ$\ell$ of 1 mol/$\ell$ $KH_2PO_4$ to 1800 m$\ell$ of distilled water, followed by 50 µ$\ell$ of 0.9 mol/$\ell$ phosphoric acid. Mobile phase: this is a solution of 215 m$\ell$ of acetonitrile in 785 m$\ell$ of phosphate buffer.

### C. Standards

Amobarbital, pentobarbital, and secobarbital can be obtained from Sigma Chemical Co., St. Louis; butabarbital and phenobarbital were obtained from the University Hospital Pharmacy; butalbital from Ganes Chemicals, New York; phenytoin from Eastman Kodak Co.,

Rochester, N.Y.; methyprylon from Hoffman-La Roche Inc., Nutley, N.J.; ethchlorvynol from Abbott Laboratories, North Chicago, Ill.; methaqualone from Ayerst Laboratories Inc., New York; glutethimide from USV Pharmaceuticals Corp., Tuckahoe, N.Y.; and 5-(4-methylphenyl)-5-phenylhydantoin, internal standard, from Aldrich Chemical Co., Milwaukee, Wis. A chromatographic standard is prepared as follows: 25 mg each of primidone, methyprylon, phenobarbital, butabarbital, butalbital, ethchlorvynol, pentobarbital, amobarbital, phenytoin, glutethimide, secobarbital, methaqualone, and 50 mg of 5-(4-methylphenyl)-5-phenylhydantoin are dissolved in 100 mℓ of methanol. This solution is stable at 4°C for at least 1 year.

### D. Procedure

Add 200 µℓ of acetonitrile containing 10 µg of the internal standard to 200 µℓ of serum in an Eppendorf® 1.5 mℓ polypropylene microtube (the method can be adapted to 25 µℓ of serum sample if desired). Vortex mix the mixture for 10 sec, then centrifuge for 1 min in an Eppendorf® #5412 centrifuge. Inject approximately 20 µℓ of the supernate onto the chromatograph and elute with the mobile phase, at a flow rate of 3.0 mℓ/min at 50°C. The column effluent is monitored at 195 nm (Figure 1).

### E. Calculation

Calculate the response factor from the chromatographic reference standard.

$$\frac{\text{peak height of Int. Std.}}{\text{peak height of drug} \times 2} = \text{Response factor (RF)}$$

Calculate the concentration of unknowns from the respective chromatograms.

$$\frac{\text{peak height of unknown drug}}{\text{peak height of Int. Std.}} \times \text{RF} \times 50 = \mu g/m\ell \text{ of unknown drug}$$

## IV. RESULTS

### A. Linearity

The method is linear up to 100.0 mg/ℓ for each drug with minimum sensitivity of 0.25 mg/ℓ for primidone and phenobarbital; 0.5 mg/ℓ for butabarbital, butalbital, phenytoin, and glutethimide, and 1.0 mg/ℓ for methyprylon, ethchlorvynol, pentobarbital, amobarbital, secobarbital, and methaqualone.[2]

### B. Recovery

Analytical recovery for these drugs ranges from 93 to 112%[2].

### C. Precision

Both within-day and day-to-day precision varies between 3.8 to 10.4% coefficient of variation.[2]

### D. Interference

Of the 35 drugs thus far tested, only ethotoin interferes with phenobarbital. At higher concentrations, mephobarbital can interfere with the accurate quantitation of amobarbital, but mephorbarbital is rapidly metabolized to phenobarbital and this is seldom detected in serum as the parent drug. Hemolyzed, lipemic, and icteric samples do not interfere with the analysis.

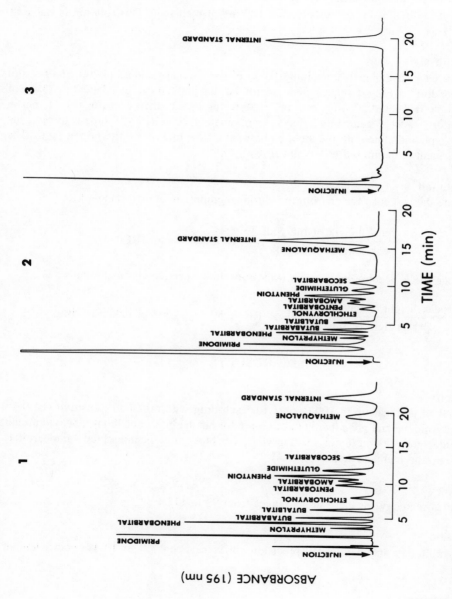

FIGURE 1. Left: chromatogram of a standard mixture of drugs. Middle: Chromatogram of a serum supplemented with 20 mg/ℓ of each drug, except for internal standard, which was 50 mg/ℓ. The chromatograph was run at 22.5% acetonitrile concentration. Right: chromatogram of a drug-free serum with 50 mg of added internal standard per liter. (From Kabra, P. M., Koo, H. Y., and Marton, L. J., *Clin. Chem.*, 24, 657, 1978. With permission.)

## V. COMMENTS

The concentration of acetonitrile in the mobile phase is critical for the resolution of these drugs. If the concentration of acetonitrile is reduced to 20.5%, the resolution between phenytoin and glutethimide is completely lost. On the other hand, if the concentration of acetonitrile is increased to 22.5%, the resolution between pentobarbital, amobarbital, and ethchlorvynol is lost. We have also noted on occasion a slight difference in selectivity between two different μBondapak® columns. Considering these factors acetonitrile concentration must be optimized for each individual column.

## REFERENCES

1. **Kabra, P. M., Koo, H. Y., and Marton, L. J.**, Hypnotics and sedatives, in *Liquid Chromatography in Clinical Analysis*, Kabra, P. M. and Marton, L. J., Eds., Humana Press, Clifton, N.J., 1981, 223.
2. **Kabra, P. M., Koo, H. Y., and Marton, L. J.**, Simultaneous liquid chromatographic determination of 12 common sedatives and hypnotics in serum, *Clin. Chem.*, 24, 657, 1978.

Chapter 16

# PENTOBARBITAL BY ULTRAVIOLET DETECTION

## Jeffrey H. Wall and George R. Gotelli

### TABLE OF CONTENTS

| | | |
|---|---|---|
| I. | Introduction | 90 |
| II. | Principle | 90 |
| III. | Materials and Methods | 90 |
| | A. Equipment | 90 |
| | B. Reagents | 90 |
| | C. Standards | 90 |
| | D. Procedure | 90 |
| | E. Calculations | 91 |
| IV. | Results | 92 |
| | A. Optimization of Chromatography | 92 |
| | B. Linearity | 92 |
| | C. Recovery | 92 |
| | D. Reproducibility | 92 |
| | E. Interferences | 92 |
| | F. Accuracy | 92 |
| V. | Comments | 92 |
| References | | 92 |

## I. INTRODUCTION

The use of high dose barbiturate induced coma to reduce intracranial pressure and brain metabolic requirements was reported by Marshall et al.[1] in 1977. Since then the use of i.v. pentobarbital has become part of the standard treatment protocol for acute brain trauma and advanced Reye's Syndrome.[2] Monitoring serum pentobarbital levels is helpful during both the loading and maintenance portions of the therapy.

The liquid chromatography (LC) method described here is simple, rapid, and specific. A single sample can be processed and chromatographed in 10 min (Figure 1).

## II. PRINCIPLE

Pentobarbital and added internal standard are extracted from serum by adsorption onto a Bond-Elut® C-10 extraction column. After serum proteins are washed off the column with distilled water, pentobarbital and the internal standard are eluted with methanol. An aliquot of the methanol is injected onto the column, detected at 210 nm, and quantitated by measuring peak height ratios.

## III. MATERIALS AND METHOD

### A. Equipment

An HPLC system equivalent to the following is recommended. A Series 1 pump and an LC-65T combination oven and variable wavelength UV detector (Perkin-Elmer Corp., Norwalk, Conn.), a Rheodyne® 7105 injection valve (Rheodyne, Cotati, Calif.), an Altex Ultrasphere® ODS column, 15 cm × 4.6 mm i.d. (Altex Scientific, Berkeley, Calif.), and a 10 mV strip-chart recorder.

### B. Reagents

Potassium dihydrogen phosphate, 1 mol/ℓ (pH 4.4) Mobile phase: 33% methanol and 14% acetonitrile in 10 mmol/ℓ phosphate buffer. Combine 330 mℓ of methanol, 140 mℓ of acetonitrile, and 10 mℓ of 1 mol/ℓ phosphate buffer with 520 mℓ of distilled water. Methanol and acetonitrile were from Burdick and Jackson Laboratories, Inc., Muskegon, Mich. Bond-Elut® extraction columns are obtained from Analytichem International, Inc., Harbor City, Calif.

### C. Standards

The stock internal standard is prepared by dissolving 25 mg of allycyclopentenyl barbituric acid (cyclopal) in 100 mℓ of distilled water. The working internal standard is a 1:10 dilution of the stock standard in 0.1 mol/ℓ phosphate buffer, (pH 4.4). Both solutions are stable at least 6 months at 4°C. The reference standard is 25 µg/mℓ of pentobarbital and 25 µg/mℓ of cyclopal in methanol. It is stable at least 1 year at 4°C. Pentobarbital and cyclopal were obtained from Applied Sciences Laboratory, Inc., State College, Pa.

### D. Procedure

When using Bond-Elut® extraction columns it is convenient to use the Vac-Elut® vacuum chamber (Analytichem International). Set up an extraction column for each sample. Pass two column volumes of methanol and two column volumes of water through each column. Disconnect the vacuum. Onto each column, pipette 200 µℓ of the working internal standard followed by 200 µℓ of sample. Connect the vacuum, then wash the columns with 2 vol of water. Label a set of 10 × 75 mm glass tubes corresponding to the samples and place them in the Vac-Elut® chamber under the appropriate columns. Pipette 300 µℓ of methanol onto

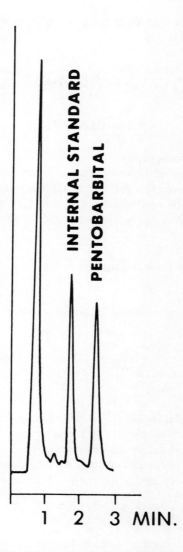

FIGURE 1. Serum containing 30 μg of pentobarbital per milliliter of blood (obtained in the authors' laboratory).

each column, connect the vacuum, and elute the columns into the collecting tubes. Inject 20 μℓ of the methanol eluate onto the column, using the following conditions: flow rate 3.0 mℓ/min; detector 210 nm, and oven temperature 50°C. The columns may be washed several times with methanol and reused at least ten times.

## E. Calculations

1. Calculate a response factor (RF) from the reference standard chromatogram as follows:

$$\frac{\text{peak height of Int. Std.}}{\text{peak height of pentobarbital}} = RF$$

2. Calculate unknown pentobarbital levels from their respective chromatograms and the RF as follows:

$$\frac{\text{peak height of pentobarbital}}{\text{peak height of Int. Std.}} \times RF \times 25 = \text{conc. of pentobarbital in } \mu g/m\ell$$

## IV. RESULTS

### A. Optimization of Chromatography
The pH and ionic strength of the mobile phase are not critical. The mobile phase was originally developed to separate five anticonvulsant drugs and separates pentobarbital from phenytoin, which is frequently coadministered to pentobarbital induced coma patients. The 50°C oven temperature improves peak resolution and stabilizes retention times.

### B. Linearity
The method is linear to at least 200 µg/mℓ.

### C. Recovery
Both absolute and analytical recovery for pentobarbital and cyclopal exceed 95%.

### D. Reproducibility
Within-run precision (CV) for a series of 20 injections was 2.7%. Day-to-day precision (CV) over a period of 1 year was 6.7%.

### E. Interferences
The method has been tested for interference from common anticonvulsants, barbiturates, benzodiazepines, and antiarrhythmics. Only carbamazepine interfered. This anticonvulsant is seldom encountered in barbiturate induced coma patients. The presence of carbamzepine can be ruled out by reinjecting patient samples with the detector wavelength set at 280 nm. Pentobarbital does not absorb at 280 nm, while carbamazepine absorbs strongly at this wavelength.

### F. Accuracy
This method has not been compared to any preexisting method.

## V. COMMENTS

This method can be used easily to analyze a single pentobarbital sample on a HPLC system which is used daily for anticonvulsant analyses. The sensitivity is more than adequate for the high blood levels (25 to 40 µg/mℓ of pentobarbital used to induce coma.

## REFERENCES

1. **Marshall, L. F., Bruce, D. A., Bruno, L., and Schut, L.,** Role of intracranial pressure monitoring and barbiturate therapy in malignant hypertension, *J. Neurol.*, 47, 481, 1977.
2. **Marshall, L. F.,** Care of acute brain injury, *Guidelines Neurosci.*, 3, 2, 1979.

Chapter 17

# CHLORAMPHENICOL BY ULTRAVIOLET DETECTION

### George R. Gotelli and Jeffrey H. Wall

## TABLE OF CONTENTS

I. Introduction ................................................................... 94

II. Principle ...................................................................... 94

III. Materials and Methods ........................................................ 94
    A. Equipment ............................................................... 94
    B. Reagents ................................................................. 94
    C. Standards ................................................................ 94
    D. Procedure ................................................................ 94
    E. Calculations ............................................................. 95

IV. Results ........................................................................ 96
    A. Optimization of Chromatography ........................................ 96
    B. Linearity ................................................................. 96
    C. Recovery ................................................................. 96
    D. Reproducibility .......................................................... 96
    E. Interference ............................................................. 96
    F. Accuracy ................................................................. 96

V. Comments ..................................................................... 96

Reference ............................................................................ 96

## I. INTRODUCTION

Chloramphenicol is a broad spectrum antibiotic active against some rickettsia and many Gram-negative and Gram-positive bacteria. Its use, however, has been associated with toxic effects, including aplastic anemia, leukopenia, thrombocytopenia, and bone marrow aplasia. To insure that therapeutic blood concentrations of chloramphenicol are achieved, while avoiding toxicity, frequent monitoring of serum levels is desirable (Figure 1).

## II. PRINCIPLE

Chloramphenicol and added internal standard are removed from serum by adsorption onto a Bond-Elut® C-10 extraction column. The column is washed with water to remove the more polar serum constituents and the drugs are eluted with methanol. An aliquot of the methanol is injected onto a reversed-phase C18 column. The separated drugs are detected at 270 nm and quantitated by peak height ratios.

## III. MATERIALS AND METHOD

### A. Equipment

A liquid chromatography (LC) system equivalent to the following is required: a Series 1 pump, an LC-100 oven, an LC-75 variable wavelength detector, and a Model 123 10 mV strip-chart recorder (Perkin-Elmer Corp., Norwalk, Conn.), a Rheodyne® 7105 injection valve, (Rheodyne, Cotati, Calif.), and an Altex Ultrasphere® C-18 column, 15 cm length × 4.6 mm i.d. (Altex Scientific, Berkeley, Calif.).

### B. Reagents

The mobile phase consists of 33% methanol and 14% acetonitrile in a 10 mmol/$\ell$ potassium dihydrogen phosphate buffer, pH 4.4. The extraction columns (Bond-Elut® columns) are purchased from Analytichem International Inc., Harbor City, Calif.

### C. Standards

Chloramphenicol and 4-chloroacetanilide (internal standard) were purchased from Aldrich Chemical Co., Milwaukee, Wis. The stock internal standard consists of 25 mg of 4-chloroacetanilide in 100 m$\ell$ of methanol. This solution is stable for at least 6 months at 4°C. The working internal standard is prepared by diluting the stock solution tenfold with water. The reference standard consists of 25 μg/m$\ell$ each of chloramphenicol and 4-chloroacetanilide in methanol.

### D. Procedure

*Note:* The Bond-Elut® extraction columns work most efficiently when used with Vac-Elut® vacuum chamber supplied by Analytichem International.

Wash the extraction columns with two column volumes of methanol and two column volumes of water. Fill each extraction column approximately 3/4 full of water and then add 200 μ$\ell$ of the working internal standard and 200 μ$\ell$ of the appropriate serum or control. Allow the extraction columns to drain completely, then pass two columns of water through each column, discarding the eluates. Place the drained extraction columns over a set of collection tubes and pass 300 μ$\ell$ of methanol through the columns collecting the eluates.

*Note:* The columns may be washed for reuse by passing several volumes of methanol through them.

Mix the contents of the collection tubes and inject a 20 μ$\ell$ aliquot onto the liquid

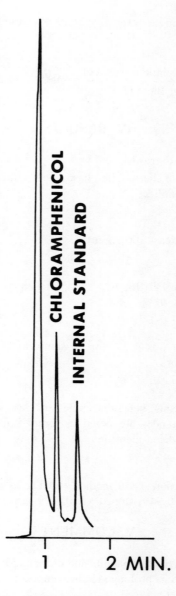

FIGURE 1. Chromatogram of a sample containing 17 μg of chloramphenicol per milliliter serum (obtained in the author's laboratory).

chromatographic column using the following instrument settings: mobile phase flow rate of 3 mℓ/min, oven temperature at 50°C, and detector set at 270 nm.

### E. Calculations

Calculate a response factor (RF) from the reference standard chromatogram as follows:

$$\frac{\text{peak height of Int. Std.}}{\text{peak height of chloramphenicol}} = RF$$

Calculate the unknowns from the respective unknown chromatograms and response factor (RF) as follows:

$$\frac{\text{peak height of unk. chloramphenicol}}{\text{peak height of Int. Std.}} \times RF \times 25 = \text{chloramphenicol conc. in } \mu g/m\ell$$

## IV. RESULTS

### A. Optimization of Chromatography
The exact organic content of the mobile phase is not critical for the separation of chloramphenicol from 4-chloroacetanilide.

### B. Linearity
The method is linear to at least 100 $\mu g/m\ell$.

### C. Recovery
Absolute recovery exceeds 95% for both chloramphenicol and 4-chloroacetanilide. Analytical recoveries exceeded >95%.

### D. Reproducibility
Within-run precision (CV) was 4.2% for 20 samples. Day-to-day precision (CV) was 10.1% over a period of 1 year.

### E. Interference
Most common anticonvulsants, benzodiazepines analgesics, and antidysrhythmics do not interfere. Occasional interferences are noted in serum samples from newborns, but the interfering substance has not been identified.

### F. Accuracy
When compared to an immunoassay method (EMIT, Syva Co., Palo Alto, Calif.), the regression analysis was r = 0.949, slope = 1.029, y-intercept = 0.613, and n = 20.

## V. COMMENTS

The half-life of chloramphenicol in adults and older children is 1.5 to 5 hr, in premature and term neonates the half-life is prolonged. Chloramphenicol is approximately 10% protein bound.

## REFERENCES

1. **Koup, J. R., Brodsky, B., Lau, A., and Beam, T. R.,** High performance liquid chromatographic assay of chloramphenicol in serum, *Antimicrob. Agents Chemother.*, 14, 439, 1978.

Chapter 18

# CHLORPROMAZINE IN PLASMA BY LIQUID CHROMATOGRAPHY/ELECTROCHEMISTRY

**Julie Morris and Ronald E. Shoup**

## TABLE OF CONTENTS

| | | |
|---|---|---|
| I. | Introduction | 98 |
| II. | Principle | 98 |
| III. | Materials and Equipment | 98 |
| | A. Equipment | 98 |
| | B. Reagents | 98 |
| | C. Standards | 98 |
| | D. Procedure | 99 |
| | E. Calculations | 99 |
| IV. | Results | 99 |
| | A. Optimization of Chromatography | 99 |
| | B. Linearity | 99 |
| | C. Recovery and Detection Limits | 99 |
| References | | 100 |

## I. INTRODUCTION

The phenothiazines have been prescribed for the treatment of psychotic patients in recent years. Chlorpromazine, an aliphatic phenothiazine, is used most frequently in the treatment of schizophrenia. It is also useful in the prevention of nausea and vomiting.

There is a strong correlation between plasma concentrations of phenothiazines and their therapeutic efficacy. However, the concentrations of phenothiazines in plasma are quite low, measured in nanograms per milliliter, and highly sensitive techniques are required to assay plasma concentrations. A GC/MS method[1] has been described which measures low plasma levels (1 to 5 ng/mℓ). However, the assay is long and tedious, and requires the synthesis of derivatives and expensive instrumentation. A GLC method[2] has been reported for the assay of seven phenothiazines; however, the detection limits reported were rather high: 50 ng/mℓ plasma. A liquid chromatography/electrochemistry (LCEC) method is described by Wallace et al.[3] for the determination of promethazine and other phenothiazines. The following outlines an analytical protocol for the determination of chlorpromazine in plasma using LCEC. It offers low detection limits and requires minimal sample preparation.

## II. PRINCIPLE

Chlorpromazine is extracted from plasma with hexane. The plasma extract is injected onto a reverse-phase column and chlorpromazine is eluted with a phosphate-tridecylamine-methanol mobile phase. Chlorpromazine is detected amperometrically and quantitated by peak heights.

## III. MATERIALS AND EQUIPMENT

### A. Equipment

A LC-154T (Bioanalytical Systems, W. Lafayette, Ind.) liquid chromatograph using a TL-5 glassy carbon working electrode and a Biophase® Octyl 5 μm column (Bioanalytical Systems, 25 cm × 4.6 mm i.d.) was used for all determinations. A nitrogen evaporator (Organomation Associates, Northborough, Mass.), a Sorvall® GLC-2 centrifuge (DuPont Company, Newton, Conn.) and a mechanical shaker (Eberbach Corporation, Ann Arbor, Mich.) were also utilized.

### B. Reagents

**Phosphate buffer (0.02 $M$)** — Add 2.7 mℓ concentrated $H_3PO_4$ (Mallinckrodt, St. Louis) to a 2 ℓ volumetric flask. Dilute to the mark with deionized, distilled water. If necessary, adjust the pH to 2.50 to 2.55 with either concentrated $H_3PO_4$ or NaOH (Mallinckrodt).

***n*-Tridecylamine (0.5 $M$)** — Dissolve in 10 g *n*-tridecylamine (Aldrich, Milwaukee, Wis.) in 100 mℓ methanol (Mallinckrodt).

**Mobile phase** — Add 5.0 mℓ 0.5 $M$ *n*-tridecylamine to a 1 ℓ volumetric flask. Add 400 mℓ methanol and mix well. Dilute to the mark with 0.02 $M$ phosphate buffer. This is the working mobile phase and should be filtered through a 0.2 μm pore size membrane (Rainin Instruments, Woburn, Mass.) and degassed prior to use. (See Table 1 for chromatographic conditions.)

**$K_2CO_3$ Buffer (0.05 $M$)** — Dissolve 6.9 g $K_2CO_3$ (Mallinckrodt) in 100 mℓ deionized, distilled water. Hexane (SpectrAR® grade, Mallinckrodt).

### C. Standards

**Chlorpromazine stock solution** — Dissolve 30 mg chlorpromazine (Sigma Chemical Co., St. Louis) in 25 mℓ mobile phase. The standard is stable at 4°C for 1 month.

## Table 1
## LIQUID CHROMATOGRAPHIC CONDITIONS FOR PLASMA CHLORPROMAZINE ASSAY

**Liquid chromatograph:** LC-154T (Bioanalytical Systems)
**Mobile phase:** 60% 0.02 $M$ $H_3PO_4$, pH 2.5/40% MeOH, containing 2.5 m$M$ $n$-tridecylamine
**Flow rate:** 1.9 m$\ell$/min
**Stationary phase:** Biophase® $C_8$ 5 μm (25 cm × 4.6 mm i.d., Bioanalytical Systems)
**Temperature:** 40°C
**Detector:** LC-4/LC-17 detector using a TL-5 glassy carbon working electrode
**Detector potential:** +0.85 V (vs. Ag/AgCl)

**Working standard solutions** — Dilute the stock solution with appropriate volumes of mobile phase.

### D. Procedure

To 1.0 m$\ell$ plasma, add 1.0 m$\ell$ 0.5 $M$ $K_2CO_3$ in 15 m$\ell$ conical centrifuge tubes. Vortex 5 to 10 sec. Add 5.0 m$\ell$ of hexane. Mechanically shake for 15 min, centrifuge at 2000 rpm for 5 min.

Transfer hexane layer to a 5 m$\ell$ test tube. Evaporate the hexane to dryness under a nitrogen steam at 40 to 45°C. Reconstitute the residue in 100 μ$\ell$ of mobile phase and inject 50 μ$\ell$. Total chromatographic time for chlorpromazine to elute is less than 6 min (Figure 1).

All test tubes, centrifuge tubes, and storage bottles were surface treated with PROSIL®-28 (PCR Research Chemicals, Gainesville, Fla., available through American Scientific Products, McGaw Park, Ill.) an organosilane surface treating agent to improve recoveries and precision.

### E. Calculations

Peak heights for unknown plasma samples are compared to spiked plasma samples whose concentrations are known. The concentration in the unknown equals the concentration of the spiked plasma times the ratio of peak heights, assuming equal volumes injected.

## IV. RESULTS

### A. Optimization of Chromatography

$n$-Tridecylamine improved peak symmetry and reduced the capacity factor. It was crucial for a sharp efficient separation, producing a dramatic effect, probably a result of an ion-repulsion mechanism. The lipophilic tridecylamine adsorbs onto the packing material and transforms it into an anion exchanger. The positive charge of chlorpromazine reduces the molecule's affinity for the column and accelerates elution. Peak symmetry is also improved.

### B. Linearity

The LCEC system was linear for injections over the range of 12 to 108 ng chlorpromazine, based on peak height measurements.

### C. Recovery and Detection Limits

The recovery of chlorpromazine was determined by comparing the current response of spiked plasma extracts to that of a standard solution of chlorpromazine. The percent recovery was calculated at a concentration of 120 ng chlorpromazine per milliliter plasma (n = 11) and was found to be 90 ± 4.8%.

The minimum detectable concentration, using a signal-to-noise ratio of 5, was determined to be 0.7 ng injected, corresponding to a concentration of 1.4 ng/m$\ell$ plasma.

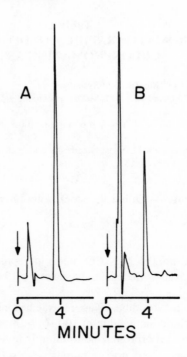

FIGURE 1. Spiked plasma extract chromatograms. (A) 54 ng chlorpromazine injected, corresponds to 120 ng/m$\ell$ plasma, 20 nAFS; (B) 6.5 ng chlorpromazine injected, corresponds to 14.4 ng/m$\ell$ plasma, 5 nAFS.

## REFERENCES

1. **Alfredson, G., Wode-Helgoadt, B., and Sedvall, G.,** A mass fragmentographic method for the determination of chlorpromazine and two of its active metabolites in human plasma and CSF, *Psychopharmacology*, 48, 123, 1976.
2. **Dinour, E., Goltschalk, L., Nandi, B., and Geddes, P.,** GLC analysis of thioridazine, mesoridazine and their metabolites, *J. Pharm. Sci.*, 65, 667, 1976.
3. **Wallace, J. E., Shimek, E. L., Stavchansky, S., and Harris, S. C.,** Determination of promethazine and other phenothiazines by liquid chromatography with electrochemical detection, *Anal. Chem.*, 53, 960, 1981.

Chapter 19

# OXAZEPAM, DIAZEPAM, AND N-DESMETHYLDIAZEPAM IN HUMAN BLOOD BY ULTRAVIOLET DETECTION

**Pokar M. Kabra**

## TABLE OF CONTENTS

I. Introduction ................................................................. 102

II. Principle ................................................................... 102

III. Materials and Methods ...................................................... 102
    A. Equipment ............................................................ 102
    B. Reagents ............................................................. 102
    C. Standards ............................................................ 102
    D. Procedure ............................................................ 103

IV. Results .................................................................... 104
    A. Calculation .......................................................... 104
    B. Linearity ............................................................ 104
    C. Recovery ............................................................. 104
    D. Precision ............................................................ 104
    E. Sensitivity .......................................................... 104
    F. Interference ......................................................... 104

V. Comments ................................................................... 104

References ..................................................................... 104

## I. INTRODUCTION

Diazepam is one of the most widely-prescribed drugs today and is present in many specimens screened for drugs.[1] Diazepam produces some sedation and relief of anxiety at low doses.[2] It is also used for the treatment of neuromuscular distress and seizures.[3] Diazepam is metabolized to N-desmethyldiazepam and oxazepam. The presence and accumulation of N-desmethyldiazepam and oxazepam is important because both of these metabolites possess significant biological activity.[4,5] Thus, monitoring the concentration of all three compounds is important.

Colorimetric,[6] spectrophotometric,[7] thin layer,[8] and gas liquid chromatography[9,10] methods have been employed in analyzing these compounds in physiological fluids. In general, these methods either do not separate diazepam and its major metabolites, are time consuming, require derivatization, or involve a long analysis time thus making the assay impractical for routine use in clinical laboratories. Liquid chromatography (LC) has advantages for the analysis of benzodiazepines in that compounds may be analyzed without an initial derivatization. Reversed-phase LC with UV detection has been widely used for the analysis of diazepam and its metabolites for therapeutic drug monitoring and pharmacokinetic studies.[11,12] The liquid chromatographic method presented here is simple, sensitive, and specific and can be used for therapeutic monitoring of these compounds.[13]

## II. PRINCIPLE

Diazepam and its metabolites, along with the internal standard prazepam, are extracted from buffered plasma or whole blood into diethylether. The extract is evaporated, reconstituted in ethanol, and an aliquot is injected onto a reversed-phase octadecyl column. The drugs are eluted with an acetonitrile-acetate buffer mobile phase and the column effluent is monitored at 240 or 254 nm (Figure 1).

## III. MATERIALS AND METHODS

### A. Equipment

A liquid chromatograph system equivalent to the following is recommended: a Model Series 1 or Series 2 liquid chromatograph equipped with a variable wavelength detector (LC 75) or a fixed wavelength detector (LC 15) (all from Perkin-Elmer Corp., Norwalk, Conn.), and a 25 cm × 4.5 mm reversed-phase column (Partisil®-10 ODS, Whatman, Inc., Clifton, N.J.) is used. The sample is injected into a Model 7105 valve (Perkin-Elmer). The column is eluted with acetonitrile/10 mmol/$\ell$ sodium acetate buffer, pH 4.6 (35/65, by vol) at a flow rate of 2.0 m$\ell$/min. The column effluent is monitored either at 240 or 254 nm.

### B. Reagents

All reagents are analytical grade. Diethyl ether AR grade can be obtained from Mallinckrodt, St. Louis. Acetonitrile, distilled in glass, UV grade is from Burdick and Jackson, Muskegon, Mich. Phosphate buffer, 1 mol/$\ell$ pH 7.0 is prepared by dissolving 136.1 g of anhydrous $KH_2PO_4$ in 1 $\ell$ of water and adjusting the pH to 7.0 with 1 $M$ $K_2HPO_4$ solution.

Acetate buffer, 10 mmol/$\ell$ is prepared by dissolving 1.42 g of sodium acetate in 1 $\ell$ of water, and adjusting the pH to 4.6 with acetic acid. The mobile phase is prepared by diluting 350 m$\ell$ of acetonitrile with 650 µ$\ell$ of acetate buffer.

### C. Standards

Diazepam and N-desmethyldiazepam are obtained from Hoffman-La Roche, Nutley, N.J.;

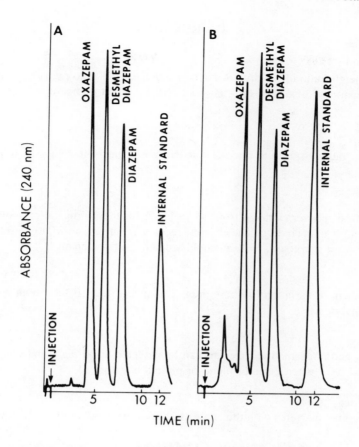

FIGURE 1. (A) Chromatogram of a standard mixture of diazepam, N-demethyldiazepam, and oxazepam (1 μg each); (B) chromatogram of a blood sample containing 1 mg/ℓ each of oxazepam, N-desmethyldiazepam, and diazepam. (From Kabra, P. M., Stevens, G. L., and Marton, L. J., *J. Chromatogr.*, 150, 355, 1978. With permission.)

oxazepam from Wyeth Laboratory, Philadelphia, Pa., and prazepam (internal standard) from Warner-Lambert Research Institute, Morris-Plains, N.J.

A chromatography standard is prepared by dissolving 10 mg each of oxazepam, N-desmethyldiazepam, diazepam, and prazepam in 100 mℓ of ethanol. A stock internal standard is prepared by dissolving 10 mg of prazepam in 1 ℓ of ethanol. A working internal standard is prepared by diluting the stock internal standard tenfold with water.

### D. Procedure

Place 2.0 mℓ of 1 $M$ phosphate buffer (pH 7.0) and 2.0 mℓ of working internal standard into a 40 mℓ centrifuge tube and then add 2.0 mℓ of whole blood, serum, or plasma. Vortex-mix each tube 5 to 10 sec, pipette 20 mℓ of anhydrous ether into each tube, shake for 5 min, and then centrifuge at 2000 rpm (210 × g) for 5 min. Shake the ether phase with 2.5 mℓ of 6 mol/ℓ HCl and centrifuge at 2000 rpm for 5 min. Aspirate the ether layer and discard, add 2.5 mℓ of 6 mol/ℓ NaOH to the aqueous phase. Then add 2.0 mℓ of 1 mol/ℓ phosphate buffer (pH 7.0) and 20 mℓ of anhydrous diethyl ether. Shake for 5 min, then centrifuge at 2000 rpm for 5 min. Aspirate the ether phase and evaporate slowly to dryness at 37°C with a stream of nitrogen. Dissolve the residue in 30 μℓ of ethanol and inject 10 to 15 μℓ onto the chromatograph.

## IV. RESULTS

### A. Calculation
The peak height ratio of the drug to internal standard is used to prepare a standard curve. The concentration of unknown drug is calculated by direct comparison with the standard curve.

### B. Linearity
Peak height ratios of drug to internal standard are linear from 50 µg to at least 10 mg/ℓ of each drug.[13]

### C. Recovery
The analytical recoveries range from 91 to 116% for these drugs at concentrations from 50 µg/ℓ to 2 mg/ℓ of each drug. The absolute recoveries range from 70 to 75% for oxazepam and 75 to 80% for diazepam, N-desmethyldiazepam, and prazepam.[13]

### D. Precision
The coefficient of variation for these drugs equals less than 10% for both within-day and day-to-day analyses.[13]

### E. Sensitivity
Oxazepam and N-desmethyldiazepam can be detected at 30 µg/ℓ and diazepam can be detected at 40 µg/ℓ when 2 mℓ of blood or plasma is extracted.[13]

### F. Interference
Carbamazepine interferes with the analysis of oxazepam and amitriptyline interferes with diazepam.[13]

## V. COMMENTS

The molarity and the pH of acetate buffer in the mobile phase is important to eliminate the peak tailing associated with the acetonitrile-water mobile phase. The variable wavelength detector can be replaced by a fixed wavelength detector. Although 254 nm could be used in most circumstances, the absorbance of benzodiazepines at this wavelength is only about half of that at 240 nm. Diazepam and N-desmethyldiazepam can be analyzed following a single solvent extraction. Chlordiazepoxide can also be analyzed by using the same mobile phase and column.

## REFERENCES

1. **Busto, U., Kaplan, H., and Sellers, E. M.**, Benzodiazepine associated emergencies in Toronto, *Am. J. Psychiatr.*, 137, 224, 1980.
2. **Cole, J. O.**, Drug treatment of anxiety, *South. Med. J.*, 71(2), 10, 1978.
3. **Gallagher, R. M.**, The impact of benzodiazepines on clinical practice: considerations for therapeutic use, *Minn. Med.*, 62, 580, 1979.
4. **Gluckman, M. I.**, Pharmacology of oxazepam (Serax), a new antianxiety agent, *Curr. Ther. Res.*, 7, 721, 1965.
5. **Randall, L. O., Schackel, C.-L., and Banziger, R.**, Pharmacology of the metabolites of chlordiazepoxide and diazepam, *Curr. Ther. Res.*, 7, 590, 1965.

6. **Baumler, J. and Rippstein, S.,** Uber den nachweis von methamino diazepoxid (librium) und seines metaboliten, *Helv. Chim. Acta,* 44, 2208, 1961.
7. **Koechim, B. A. and D'Arconte, L.,** Determination of chlordiazepoxide (librium) and of a metabolite of lactam character in plasma of humans, dogs and rats by a specific spectrofluorometric micro method, *Anal. Biochem.,* 5, 195, 1963.
8. **Vam-der Merwe, P. J. and Steyn, J. M.,** Thin layer chromatographic method for determination of diazepam and its major metabolites, *J. Chromatogr.,* 148, 549, 1978.
9. **De Silva, J. A. F., Bekersky, I., Puglisi, C. V., Brooks, M. A., and Weinfeld, R. E.,** Determinations of 1,4-benzodiazepines and diazepin-2-ones in blood by electron-capture gas-liquid chromatography, *Anal. Chem.,* 48, 10, 1976.
10. **De Silva, J. A. F. and Puglisi, C. V.,** Determination of medazepam (nobrium), diazepam (valium) and their major biotransformation products in blood and urine by electron capture gas-liquid chromatography, *Anal. Chem.,* 42, 1725, 1970.
11. **MacKichan, J. J., Jusko, W. J., Duffner, P. K., and Cohen, M. E.,** Liquid-chromatographic assay of diazepam and its major metabolites in plasma, *Clin. Chem.,* 25, 856, 1979.
12. **Cotler, S., Puglisi, C. V., and Gustafson, J. H.,** Determination of diazepam and its major metabolites in man and cat by high-performance liquid chromatography, *J. Chromatogr.,* 222, 95, 1981.
13. **Kabra, P. M., Stevens, G. L., and Marton, L. J.,** High-pressure liquid chromatographic analysis of diazepam, oxazepam, and N-desmethyldiazepam in human blood, *J. Chromatogr.,* 150, 355, 1978.

Chapter 20

# CHLORTHALIDONE IN BLOOD AND URINE

## Emil T. Lin

## TABLE OF CONTENTS

I. Introduction ................................................................. 108

II. Principle ..................................................................... 108

III. Materials and Methods .................................................... 108
    A. Equipment ............................................................. 108
    B. Reagents ............................................................... 108
    C. Standards .............................................................. 108
    D. Procedure ............................................................. 109
    E. Calculation ............................................................ 109

IV. Results ...................................................................... 109
    A. Linearity ............................................................... 109
    B. Recovery .............................................................. 110
    C. Interference .......................................................... 110
    D. Precision .............................................................. 110
    E. Metabolites .......................................................... 110

V. Comments .................................................................. 110

References ........................................................................ 110

## I. INTRODUCTION

Chlorthalidone is a diuretic with pharmacological effects similar to thiazides, although it differs chemically from the thiazide group. It has a relative long half-life (~45 hr), correspondingly long duration of effect, and a low incidence of adverse effects. Previous gas chromatographic assay procedures for the analyses of chlorthalidone in blood and urine[1,2] are based on the conversion of chlorthalidone to its tetramethyl derivative by extractive alkylation. The methods are tedious and may pose methodological problems, especially when analyzing a large number of samples. A HPLC method for determination of chlorthalidone in whole blood and urine samples was developed.[3] The method is simple, rapid, sensitive, and requires 0.2 mℓ of biological fluid.

## II. PRINCIPLE

Whole blood is mixed with an equal volume of water, sonicated, and then mixed with acetonitrile to precipitate the blood proteins. The acetonitrile supernatant is evaporated to one third of the original volume, separated on a reverse-phase column, and eluted with an acetonitrile-phosphate buffer mobile phase. The drug is detected by its absorption at 210 nm for blood and 250 nm for urine and quantitated from its peak height ratio of chlorthalidone to internal standard.

## III. MATERIALS AND METHODS

### A. Equipment

Liquid chromatography (LC) systems equivalent to the following should be used for both blood and urine samples: a model series 3 pump (Perkin-Elmer Corp., Norwalk, Conn.), equipped with an auto-sampler WISP® 710B (Waters Associates, Milford, Mass.), a variable wavelength spectrophotometric detector model SP770 (Kratos Analytical Instruments, Ramsey, N.J.), a Linear® dual pen charter recorder (ISI, Inc, Concord, Calif.), and a reverse-phase μBondapak® phenyl column, 10 μm, 0.4 × 30 cm (Waters Associates).

### B. Reagents

Acetonitrile, HPLC grade, (J. T. Baker, Phillipsburg, N.J.), and phosphoric acid, certified ACS grade, (Fisher Scientific, Fairlawn, N.J.). Mobile phase: 18% $CH_3CN$ in 0.1% $H_3PO_4$, adjusted to pH 2.75 with 10% NaOH. The solution is filtered through Whatman® #2 filter paper (VMR Scientific, San Francisco) and degassed under vacuum.

### C. Standards

Chlorthalidone was supplied by Mylan Pharmaceuticals (Morgantown, W.Va.), phentolamine HCl can be obtained from Ciba-Geigy (Summit, N.H.), and sodium pentobarbital was obtained from Abbott Laboratory (North Chicago, Ill.). The chlorthalidone stock standard is prepared by dissolving 4 mg of chlorthalidone into 100 mℓ of methanol. The solution is stable at 4°C for at least 6 months. The stock internal standard of blood is made by dissolving 18 mg of phentolamine HCl in 100 mℓ of water. A 100-fold dilution of the stock with acetonitrile is used as the working internal standard for blood samples. A stock internal standard for urine is made by dissolving 20 mg of sodium pentobarbital in 100 mℓ of water. A 100-fold dilution of the stock with acetonitrile is used as the working internal standard for urine samples. Standard curve of blood is prepared by spiking 0 to 25 μℓ of chlorthalidone stock solution into 0.2 mℓ of blank blood to make a standard curve of blood ranging from 0 to 5 μg/mℓ. Urine standard curve is prepared by spiking 0 to 50 μℓ of chlorthalidone

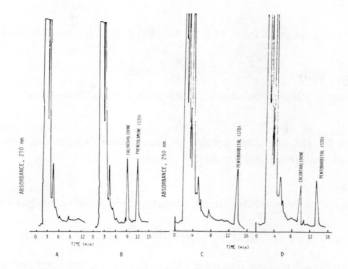

FIGURE 1. Chromatograms of (A) blood blank, (B) blood standard containing 1.02 µg/mℓ of chlorthalidone, (C) urine blank, and (D) urine standard containing 9.51 µg/mℓ of chlorthalidone.

stock solution into 0.2 mℓ of blank urine to make a standard curve of urine ranging from 0 to 10 µg/mℓ.

Quality control specimens are prepared in-house in the same manner as the blood or urine standard where low, medium, and high concentrations are prepared and stored at −30°C until assay.

### D. Procedure

Blood sample: whole blood (0.2 mℓ) is mixed with an equal volume of distilled water, sonicated for 5 min, and mixed with 0.5 mℓ of blood working internal standard solution. After mixing and centrifugation, the supernatant is transferred to a clean tube and evaporated under nitrogen until 0.4 mℓ of solution remains. Inject one third to one fifth of the sample onto the column and elute with the mobile phase at a flow rate of 2.0 mℓ/min. Detector sensitivity is set at 0.02 A full scale. The output of the dual pen recorder is set at 10 and 5 mV. The retention times for chlorthalidone and phetolamine are 10 and 12 min, respectively (Figure 1).

Urine samples are handled as blood with the exceptions that the sonication and centrifugation steps are not included and the urine internal standard is sodium pentobarbital. The retention times for chlorthalidone and pentobarbital are 10 and 15 min, respectively.

### E. Calculation

Calibration graphs are constructed from spiked blood and urine samples using the same procedure described above. The peak height ratios (drug:internal standard) are plotted vs. drug concentration in µg/mℓ and the calibration graph is used for the calculation of blood or urine concentration in human subjects.

## IV. RESULTS

### A. Linearity

Peak height ratios of chlorthalidone to internal standard are linearly related from 0.2 to at least 5.0 µg/mℓ for blood and 4 to at least 190 µg/mℓ for urine.

## B. Recovery
Absolute recovery range is from 87.0 to 108.0% in the above concentration range for blood.

## C. Interference
No interference is found from other diuretics with the condition stated.

## D. Precision
The precision for chlorthalidone equals less than 6.0% (CV) for blood and 8.0% (CV) for urine for both within-day and day-to-day analyses.

## E. Metabolites
The major part of an absorbed dose of chlorthalidone is excreted unchanged via the kidney.[4]

## V. COMMENTS

The therapeutic concentration range for chlorthalidone has not been defined. Oral dose of 50 mg chlorthalidone results in a peak blood level of about 4.62 µg/mℓ at 12 to 16 hr after dose.[5]

## REFERENCES

1. **Tweeddale, M. G. and Ogilvie, R. I.,** Improved method for estimating chlorthalidone in body fluids, *J. Pharm. Sci.,* 63, 1065, 1974.
2. **Fleuren, H. L. and Van Rossum, J. M.,** Determination of chlorthalidone in plasma, urine and red blood cells by gas chromatography with nitrogen detection, *J. Chromatogr.,* 152, 41, 1978.
3. **Peng, C-T., Lin, E. T., and Benet, L. Z.,** Reversed phase HPLC analysis of chlorthalidone in whole blood and urine, *Abstr. Pharm. Assoc.,* 11, 128, 1981.
4. **Fleuren, H. J. L., Thien, Th.A., Veuvey-van Wissen, G. P. W., and von Rossum, J. M.,** Absolute bioavailability of chlorthalidone in man, a cross over study after intravenous and oral administration, *Eur. J. Clin. Pharmacol.,* 15, 35, 1979.
5. **Williams, R., Blume, C. D., Lin, E. T., Holford, N. H. G., and Benet, L. Z.,** Relative bioavailability of chlorthalidone in humans: adverse influence of polyethylene glycol, *J. Pharm. Sci.,* 71, 533, 1982.

Chapter 21

# FUROSEMIDE IN PLASMA AND URINE

## Emil. T. Lin

## TABLE OF CONTENTS

I. Introduction .................................................................... 112

II. Principle ........................................................................ 112

III. Materials and Methods ...................................................... 112
    A. Equipment ............................................................... 112
    B. Reagents .................................................................. 112
    C. Standards ................................................................. 112
    D. Procedure ................................................................ 113
    E. Calculation .............................................................. 113

IV. Results ......................................................................... 114
    A. Linearity ................................................................. 114
    B. Recovery ................................................................. 114
    C. Interference ............................................................. 114

V. Comments ..................................................................... 114

References ............................................................................ 114

## I. INTRODUCTION

Furosemide, an anthranilic acid derivative, is one of the most potent diuretics available. It inhibits the active reabsorption of chloride in the diluting segment of the loop of Henle, thus preventing the reabsorption of sodium which passively follows chloride.[1] In contrast to the thiazides, furosemide is effective during renal failure. Clinically used doses range from 20 to 3000 mg. The earliest methods for analysis of furosemide were based on spectrophotometry[2] or spectrophotofluorometry[3] following extraction from serum and urine with organic solvents. In order to improve the specificity, HPLC methods have subsequently been developed. The liquid chromatography (LC) method described is rapid, sensitive, and accurate for quantitating furosemide in plasma and urine. By using this method, we found no evidence of the proposed metabolite of furosemide, 2-amino-4-chloro-5-sulfamoylanthranilic acid (CSA) in plasma or urine samples. In addition, we conclusively demonstrated that CSA is an analytical artifact.[4]

## II. PRINCIPLE

Furosemide in plasma (or urine) is quantitatively determined after protein precipitation with acetonitrile. The acetonitrile supernatant is evaporated to one third of the original volume and furosemide is separated on a reverse-phase column with an acetonitrile-phosphate buffer mobile phase and detected by fluorescence. Excitation and emission wavelengths are set at 345 and 405 nm, respectively. The internal standard, phenobarbital, is monitored by UV detection at 254 nm. The concentration of furosemide is quantitated from the peak height ratio of furosemide to internal standard.

## III. MATERIALS AND METHODS

### A. Equipment

Liquid chromatography systems equivalent to the following should be used for both plasma and urine samples: a Perkin-Elmer Model 3 pump (Perkin-Elmer Corp., Norwalk, Conn.) equipped with a Waters® intelligent sample processor Model 710B (Waters Associates, Milford, Mass.), a fluorescence spectrophotometer Model 204A and UV detector Model 250, (Perkin-Elmer), a Linear® dual pen charter recorder (ISI, Inc., Concord, Calif.), and a Waters® C18 μBondapak® 30 cm × 4 mm i.d. (Waters Associates).

### B. Reagents

Acetonitrile, HPLC grade, J. T. Baker, Phillipsburg, N.J.), and phosphoric acid, certified A.C.S. (Fisher Scientific, Fairlawn, N.J.). Mobile phase: 30% acetonitrile in water containing 0.1% phosphoric acid. The solvent is filtered through Whatman® #2 paper (VWR Scientific Inc., San Francisco) and degassed under vacuum.

### C. Standards

Furosemide may be obtained from Hoechst-Roussel Pharmaceuticals Inc., Somerville, N.J. Sodium phenobarbital was from Merck Sharp and Dohme Co., West Point, Pa.

Furosemide (4.1 mg) is dissolved in 100 mℓ of acetonitrile to yield a stock solution of 41 μg/mℓ. This stock solution is then diluted 5-fold (8.12 μg/mℓ) and 100-fold (0.41 μg/mℓ) to give the working standard solutions.

Appropriate volumes of furosemide working standard solution are spiked into 200 μℓ of blank plasma or urine to the concentration range of 8.2 to 205 ng/mℓ for plasma and 0.4 to 20.2 μg/mℓ for urine.

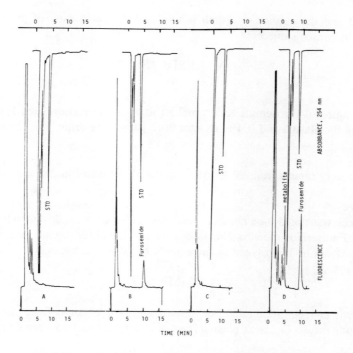

FIGURE 1. Chromatograms of (A) plasma blank containing internal standard, (B) plasma standard containing 15 ng/mℓ of furosemide, (C) urine blank with internal standard, and (D) patient urine sample containing 8.3 μg/mℓ of furosemide.

A stock internal standard is made by dissolving 1 g of sodium phenobarbital into 100 mℓ of water. A working internal standard is prepared by taking 2.5 mℓ of the stock standard and diluting to 40 mℓ with acetonitrile.

Quality control specimens are prepared in-house in the same manner as the plasma or urine standard. A proper low, medium, and high concentration control specimen is assayed for each study.

### D. Procedure

Plasma samples: to 100 μℓ of plasma sample, plasma standard, or control specimen in a culture test tube, add 300 μℓ of working internal standard. Shake the mixture on a vortex mixer and then centrifuge for 10 min. Transfer the supernatant to a clean test tube and evaporate under $N_2$ until approximately 0.1 mℓ of the solution remains. Inject 20 to 40 μℓ of the solution onto the chromatograph and elute with mobile phase at a flow rate of 2.0 mℓ/min.

Urine samples are prepared similarly except without the precipitation step. The fluorescence detector is set to a photomultiplier (PM) gain of 3 and sensitivity of 10 for plasma samples and a PM gain of 2 and sensitivity of 10 for urine samples. The UV detector is set at an attenuation of 5 and an O.D. range of 0.2 for both plasma and urine. The retention times for phenobarbital and furosemide are 4 and 9 min, respectively (Figure 1).

### E. Calculation

Plasma and urine standards are carried through the same procedure described above. The peak height ratios (furosemide/internal standard) are plotted vs. drug concentration in ng/

mℓ (or μg/mℓ) and the calibration graph is used for the calculation of plasma or urine concentrations in human subjects.

## IV. RESULTS

### A. Linearity
Peak height ratios of furosemide to internal standard are linearly related from 8.5 to at least 205 ng/mℓ for plasma and 0.4 to at least 20.2 μg/mℓ for urine.

### B. Recovery
Absolute recovery ranges from 90 to 100% in the above concentrations.

### C. Interference
No interference was noted for a number of patients taking furosemide due to the selectivity of the detector. Furosemide glucuronate, a major metabolite of furosemide in urine, is eluted within 3 min and can be hydrolyzed to furosemide by μ-glucuronidase.[4]

## V. COMMENTS

The therapeutic concentration range for furosemide has not been defined. Oral dosage of 40 mg furosemide results in a peak plasma level of about 1.17 to 3.35 μg/mℓ approximately 1 hr after dose.

## REFERENCES

1. **Jacobson, H. R. and Kokko, J. P.,** Diuretics: sites and mechanisms of action, *Ann. Rev. Pharmacol. Toxicol.,* 16, 201, 1976.
2. **Hajdu, V. P. and Haussler, A.,** Untersuchungen mit dem Salidiureticum 4-chlor-N-(furoylmethyl)-5-sulfamyl-anthranilsaure, I., *Arzneim.-Forsch.,* 14, 709, 1964.
3. **Forre, A. W., Kimpel, B., Blair, A. D., and Cutler, R. E.,** Furosemide concentrations in serum and urine, and its binding by serum proteins as measured fluorometrically, *Clin. Chem.,* 20, 152, 1974.
4. **Smith, D. E., Lin, E. T., and Benet, L. Z.,** Absorption and disposition of furosemide in healthy volunteers, measured with a metabolite-specific assay, *Am. Pharmacol. Exp. Ther.,* 18, 337, 1980.

Chapter 22

# HYDROCHLOROTHIAZIDE IN PLASMA AND URINE

**Emil T. Lin**

## TABLE OF CONTENTS

I. Introduction ................................................................. 116

II. Principle ..................................................................... 116

III. Materials and Methods ................................................. 116
    A. Equipment ............................................................ 116
    B. Reagents .............................................................. 116
    C. Standards ............................................................. 117
    D. Procedure ............................................................. 118
    E. Calculation ........................................................... 118

IV. Results ....................................................................... 118
    A. Linearity .............................................................. 118
    B. Recovery ............................................................. 118
    C. Interference .......................................................... 118
    D. Precision ............................................................. 118
    E. Metabolites .......................................................... 118

V. Comments .................................................................. 118

References ........................................................................ 118

## I. INTRODUCTION

The benzothiadiazide (thiazide) diuretics are widely used for the treatment of hypertension and fluid retention associated with congestive heart failure. Several methods have been developed to measure thiazide diuretics qualitatively or quantitatively in biological fluids. Colorimetric determination of thiazides lacks sensitivity and specificity.[1,2] A simpler spectrophotometric procedure[3] was also found to lack specificity. Liquid chromatography (LC) methods offer simplicity and specificity. Two methods have been described for the detection of thiazide diuretics in urine.[4,5] Both methods require either multiple mobile phases or different types of columns for determining the concentration of thiazides in urine. Since the various thiazide diuretics are prescribed over a large range of dosages and not formulated in a combination form, it is more practical to measure a specific thiazide utilizing a LC method specific for that particular thiazide. The method for hydrochlorothiazide described offers simplicity, sensitivity and specificity for both plasma and urine and can be easily modified for other thiazides.[6]

## II. PRINCIPLE

Plasma proteins are precipitated with acetonitrile containing bromohydrochlorothiazide (plasma) or hydroflumethiazide (urine) as an internal standard. After the mixture is centrifuged, the supernatant is evaporated to 0.1 m$\ell$ and injected onto a reverse-phase column. The drug is eluted with an acetonitrile-acetate buffer mobile phase. The drug is detected by its absorption of 271 nm and quantified from the peak height ratio of hydrochlorothiazide to internal standard.

## III. MATERIALS AND METHODS

### A. Equipment

Liquid chromatography systems equivalent to the following should be used: a model series 3 pump (Perkin-Elmer Corp., Norwalk, Conn.) equipped with a Waters® intelligent sample processor Model 710B (Waters Associates, Milford, Mass.), a variable wavelength detector Model 65T (Perkin-Elmer), and a Curken® dual pen recorder (Tegal Scientific Inc., Martinez, Calif.). An Alltech® C18-10 μm column 0.46 × 25 cm, (Alltech Associates, Deerfield, Ill.) is needed for urine and an Alltech® C8-5 μm column 0.46 × 25 cm is used for plasma sample.

### B. Reagents

Plasma sample: acetonitrile (HPLC grade) and glacial acetic acid (reagent grade) are both from J. T. Baker, Phillipsburg, N.J. Mobile phases are 15% $CH_3CN$ in 0.1% $CH_3COOH$ and 11% $CH_3CN$ in 0.1% $CH_3COOH$ for urine and plasma, respectively. The mobile phases are filtered through Whatman® #2 filter paper (VMR Scientific, San Francisco) and degassed under vacuum.

### C. Standards

Hydrochlorothiazide and bromohydrochlorothiade can be obtained from Merck Sharp and Dohme Co., West Point, Pa. Hydroflumethiazide can be obtained from Bristol Laboratories, Syracuse, N.Y.

The hydrochlorothiazide urine stock standard is prepared as follows: 38 mg of hydrochlorothiazide is dissolved in 100 m$\ell$ of methanol. Plasma stock standard is prepared by diluting the urine stock standard 100 times with methanol. The solution is stable at 4°C for at least 6 months.

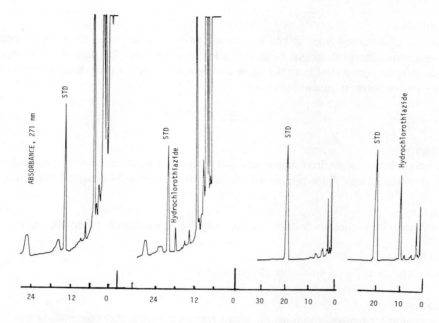

FIGURE 1. Chromatograms of (A) plasma blank containing internal standard, (B) plasma standard containing 90 ng/mℓ of hydrochlorothiazide, (C) urine blank containing internal standard, and (D) urine standard containing 28.5 μg/mℓ of hydrochlorothiazide.

The bromohydrochlorothiazide stock internal standard for plasma is made by dissolving 29 mg of bromohydrochlorothiazide in 100 mℓ of methanol. A 1000-fold dilution of the stock with acetonitrile is used as the working internal standard for plasma.

A hydroflumethiazide working internal standard for urine is made by dissolving 39.65 mg of hydroflumethiazide in 500 mℓ of acetonitrile.

A standard curve for plasma (range 0 to 195.0 ng/mℓ) is prepared by spiking 0.2 mℓ of blank plasma with 0 to 10 μℓ of plasma working standard. A urine standard curve is prepared by spiking 0.2 mℓ of blank urine with 0 to 30 μℓ of hydrochlorothiazide urine working standard (range of 0 to 57 μg/mℓ). Quality control specimens are prepared in-house in the same manner as the plasma or urine standards where low, medium, and high concentrations are prepared and stored at −30°C until assay.

## D. Procedure

Plasma sample: to 200 μℓ of plasma sample, plasma standard, or control in a culture tube, add 400 μℓ of plasma working internal standard solution. Vortex mix for 10 min, transfer the supernatants, and evaporate to 200 μℓ under a stream of nitrogen gas. Inject 10 to 20 μℓ of the sample onto the chromatograph and elute with the mobile phase at a flow rate of 0.8 mℓ/min. The column temperature is maintained at room temperature. Detector sensitivity is set at 0.004 A full scale. The output of the dual pen recorder is set at 10 and 5 mV. The retention times for hydrochlorothiazide and bromohydrochlorothiazide are 15.0 and 20.0 min, respectively (Figure 1).

Urine sample: to 200μℓ of urine sample, urine standard, and urine control in a culture tube, add 200 μℓ of urine working internal standard solution. Mix well and transfer to WISP® vial, inject 5 to 10 μℓ of the resulting sample onto the column, and elute with the mobile phase at a flow rate of 1.7 mℓ/min. Detector sensitivity is set at 0.01 A full scale. The output of the dual pen recorder is set at 10 and 5 mV. The retention times for hydrochlorothiazide and hydroflumethiazide are 10.0 and 20.0 min, respectively (Figure 1).

### E. Calculation

Graphs are constructed from spiked plasma and urine samples using the same procedure described above. The peak height ratios (drug:internal standard) are plotted vs. drug concentration in ng/mℓ (or μg/mℓ) and the calibration graph is used for the calculation of plasma or urine concentration in human subjects.

## IV. RESULTS

### A. Linearity

Peak height ratios of hydrochlorothiazide to internal standard are linearly related from 19.5 to at least 195 ng/mℓ for plasma and from 1.90 to at least 57.0 μg/mℓ for urine.

### B. Recovery

Absolute recovery range is from 90 to 100% in the concentration range above.

### C. Interference

No interference is found from other thiazides currently in use.

### D. Precision

The precision for hydrochlorothiazide equals less than 10.0% (CV) for plasma and 3.0% (CV) for urine for both within-day and day-to-day analyses.

### E. Metabolites

Hydrochlorothiazide is eliminated unchanged via the kidney.

## V. COMMENTS

The therapeutic concentration range of hydrochlorothiazide has been defined. Oral doses of 50 mg hydrochlorothiazide result in peak plasma level of about 230 ng/m/ℓ.[7]

## REFERENCES

1. **Sheppard, H., Mowles, T. F., and Plummer, A.J.**, Determination of hydrochlorothiazide in urine, *J. Am. Assoc. Ed.*, 49, 722, 1960.
2. **Resetarits, D. E. and Bates, T. R.**, Errors in chlorothiazide bioavailability estimates based on a Bratton-Marshall colorimetric method for chlorothiazide in urine, *J. Pharm. Sci.*, 68, 126, 1979.
3. **Pilsbury, V. B. and Jackson, J. V.**, Identification of the thiazide diuretic drugs, *J. Pharm. Pharmacol.*, 18, 713, 1966.
4. **Tisdall, P. A., Moyer, T. P., and Anhalt, J. P.**, Liquid chromatographic detection of thiazide diuretic in urine, *Clin. Chem.*, 26, 702, 1980.
5. **Shah, V. P., Lee, J., and Prasad, V. K.**, Thiazides. XII. A simple HPLC method for determination of thiazides in urine, *Anal. Lett.*, 15(B6), 529, 1982.
6. **Lin, E. T. and Benet, L. Z.**, High pressure liquid chromatographic determination of chlorothiazide and hydrochlorothiazide in human serum and urine, *Abstr. Am. Pharm. Assoc.*, 8, 194, 1978.
7. **Williams, R. L., Davies, R. O., Berman, R. S., Holmes, G. I., Huber, P., Gee, W. L., Lin, E. T., and Benet, L. Z.**, Hydrochlorothiazide pharmacokinetics and pharmacologic effect: the influence of indomethacin, *J. Clin. Pharmacol.*, 22, 32, 1982.

Chapter 23

# METHYCLOTHIAZIDE IN BLOOD AND URINE

## Emil T. Lin

## TABLE OF CONTENTS

I. Introduction ................................................................. 120

II. Principle ..................................................................... 120

III. Materials and Methods .................................................. 120
    A. Equipment ............................................................ 120
    B. Reagents ............................................................... 120
    C. Standards .............................................................. 120
    D. Procedure ............................................................. 120
    E. Calculation ........................................................... 121

IV. Results ....................................................................... 121
    A. Linearity ............................................................... 121
    B. Recovery .............................................................. 122
    C. Interference .......................................................... 122
    D. Precision .............................................................. 122
    E. Metabolites .......................................................... 122

V. Comments ................................................................... 122

References ........................................................................... 122

## I. INTRODUCTION

The oral diuretic-antihypertensive agent, methyclothiazide, is a member of the thiazide family of drugs that has a relatively long duration of action ($T^{1}/_{2} > 15$ hr). Unlike hydrochlorothiazide and chlorothiazides, methyclothiazide concentrates more than 30-fold in the red blood cells over plasma, thus it is preferable to assay the drug in whole blood. A simple and reproducible method for the determination of the drug level in human blood and urine is described.[1]

## II. PRINCIPLE

The drug is extracted from blood or urine with ethyl ether after sonication with an equal volume of water. The ether layer is evaporated to dryness and the residue is dissolved in acetonitrile-phosphate buffer mobile phase and injected onto the column. The drug is detected by absorption spectrophotometry and quantitated from its peak height ratio vs. internal standard.

## III. MATERIALS AND METHODS

### A. Equipment

Liquid chromatograph (LC) systems equivalent to the following should be used: a model series 3 pump (Perkin-Elmer Corp., Norwalk, Conn.) equipped with an autosample injector 710B (Waters Associates, Milford, Mass.), a variable wavelength detector, LC-65T (Perkin-Elmer), and a Linear® dual pen recorder (ISI, Inc., Concord, Calif.). A RP-18, 5 μm column, 0.46 × 25 cm (Alltech Associates, Deerfield, Ill.) is used for both plasma and urine.

### B. Reagents

Acetonitrile, HPLC grade, (J. T. Baker, Phillipsburg, N.J.), phosphoric acid, certified A.C.S. (Ficher Scientific, Fairlawn, N.J.), and sodium hydroxide, AR grade, (Mallinckrodt, Inc., St. Louis). Mobile phase is 29% $CH_3CN$ in 0.2% $H_3PO_4$ for both blood and urine. The mobile phase is filtered through Whitman® #2 filter paper (VWR Scientific, San Francisco) and degassed under vacuum.

### C. Standards

Methyclothiazide can be obtained from Abbott Pharmaceuticals, North Chicago, Ill., and prednisone from Sigma Chemical Co., St. Louis.

The methyclothiazide working standard is prepared as follows: 10 mg of methyclothiazide is dissolved into 500 mℓ of methanol. The solution is stable at 4°C for at least 6 months. The stock internal standard is made by dissolving 11.4 mg of prednisone in 100 mℓ of methanol. A tenfold dilution of the stock with 50% methanol is used as the working internal standard.

A spiked volume of 0, 1, 2, 3, 5, 10, 20, and 40 μℓ of methyclothiazide stock solution is added to 1 mℓ of urine or blood blank in culture tubes to yield concentrations of 0, 20, 40, 60, 100, 200, 400, and 800 ng/mℓ of methyclothiazide.

Quality control specimens are prepared in-house in the same manner as the blood or urine standard. Proper low, medium, and high concentration control specimens are prepared and stored at −30°C until assay.

### D. Procedure

To 1 mℓ of blood sample, blood standard, or control specimen, add 100 μℓ of working

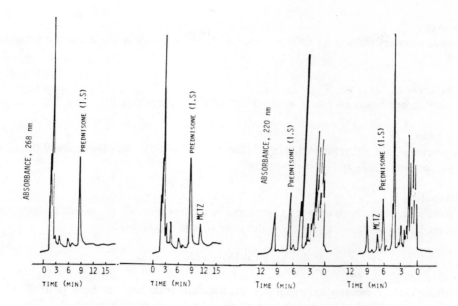

FIGURE 1. Chromatograms of (A) blood bank containing internal standard, (B) blood standard containing 63 ng/mℓ of methyclothiazide, (C) urine blank containing internal standard, and (D) urine standard containing 63 ng/mℓ of methyclothiazide.

internal standard and 1 mℓ of water and sonicate for 5 min. Extract twice with 5 mℓ of ether for 15 min using a mechanical shaker. Evaporate the organic layer to dryness under a stream of nitrogen gas. Dissolve the residue in 250 μℓ of mobile phase, inject approximately 20 μℓ of the sample onto the column, and elute with the mobile phase at a flow rate of 0.8 mℓ/min. The column temperature is maintained at room temperature and column effluent is monitored at 268 nm. Detector sensitivity is set to 0.04 A full scale. The output of the dual pen recorder is set at 1 and 2 mV. The retention times for methyclothiazide and prednisone are 21 and 15 min, respectively (Figure 1).

Urine samples are prepared similarly except the sonication step is omitted and ether extraction is performed only once. For urine, the flow rate of the mobile phase is 1.0 mℓ/min and the column effluent is monitored at 220 nm. Detector sensitivity is set to 0.04 A full scale. The output of the dual pen recorder is set at 1 and 2 mV. The retention times for methyclothiazide and prednisone are 16 and 13 min, respectively (Figure 1).

### E. Calculation

Graphs are constructed from spiked plasma and urine samples using the same procedure described above. The peak height ratio (methyclothiazide:internal standard) is plotted vs. drug concentration in ng/mℓ and the calibration graph is used for the calculation of blood or urine concentration in human subjects.

## IV. RESULTS

### A. Linearity

Peak height ratios of drug to internal standard are linearly related from 21 to 840 ng/mℓ for blood and 21 to 421 ng/mℓ for urine in actual studies.

## B. Recovery

Absolute recovery range is 91.0% for blood and 86.0% for urine in the above concentration range.

## C. Interference

No interference has been found from other thiazides currently in use.

## D. Precision

The precision for methyclothiazide equals less than 7.0% (CV) for plasma and 9.0% (CV) for urine for both intra- and inter-day.

## E. Metabolites

No metabolites have been reported in the literature. However, only 6% of methyclothiazide is recovered from urine after 10 mg of oral dose of methyclothiazide.[2]

# V. COMMENTS

The therapeutic concentration range for methyclorthiazide has not been defined. Oral doses of 10 mg methyclothiazide result in a peak blood level of 600 ng/m$\ell$ at 3.5 hr.

# REFERENCES

1. **El-Sayed, N., Gee, W. L., Lin, E. T., and Benet., L. Z.,** Methyclothiazide HPLC analysis in whole blood and urine, *Abstr. Am. Pharm. Assoc.*, 12, 160, 1982.
2. **Gee, W. L., El-Sayed, N., Lin, E. T., Blume, C., Williams, R., and Benet, L. Z.,** Bioequivalency of methyclothiazide tablet formulations in man, *Abstr. Am. Pharm. Assoc.*, 13, 53, 1983.

Chapter 24

# TRIAMTERENE AND ITS METABOLITE IN PLASMA AND URINE

### Emil T. Lin

## TABLE OF CONTENTS

I. Introduction ............................................................. 124

II. Principle ............................................................... 124

III. Materials and Methods .................................................. 124
    A. Equipment ......................................................... 124
    B. Reagents .......................................................... 124
    C. Standards ......................................................... 124
    D. Procedure ......................................................... 125
    E. Calculation ....................................................... 125

IV. Results ................................................................ 125
    A. Linearity ......................................................... 125
    B. Recovery .......................................................... 125
    C. Interference ...................................................... 125
    D. Precision ......................................................... 125
    E. Metabolites ....................................................... 126

V. Comments ............................................................... 126

References ................................................................. 127

## I. INTRODUCTION

Triamterene is a potassium-sparing diuretic used mainly in combination with potassium-wasting diuretics such as furosemide and hydrochlorothiazide to increase natriuresis and reduce kaliuresis.[1,2] The phase II metabolite of triamterene, hydroxytriamterene sulfuric acid ester, also has a diuretic effect in rats.[3] We reported the first HPLC assay[4] by which the specific measurement of hydroxytriameterene-sulfate can be obtained and by which very low concentrations of unchanged triamterene and hydroxytriamterene-sulfate can be quantitated. The method is simple, fast, and accurate.

## II. PRINCIPLE

Plasma proteins are precipitated from 200 µℓ of plasma with acetonitrile containing hydroflumethiazide as an internal standard. After centrifugation, an aliquot of the supernatant is injected onto a reverse-phase column and the drug is eluted with an acetonitrile-phosphate buffer mobile phase. Urine samples are handled similarly except omitting the precipitation step. The drug is detected by fluorescence and quantitated from the peak height ratio.

## III. MATERIALS AND METHODS

### A. Equipment

Liquid chromatography (LC) systems equivalent to the following should be used: a Beckman® pump Model 110A (Beckman Instruments, Mountain View, Calif.) equipped with a Waters® intelligent sample processor (WISP®) Model 710B (Water Associates, Milford, Mass.), a Perkin-Elmer fluorescence spectrophotometer 204S (Perkin-Elmer Corp., Norwalk, Conn.), a Linear® dual pen charter recorder (ISI, Inc., Concord, Calif.), and a reverse-phase octadecylsilane column C18, 10 µm, 0.46 × 25 cm, (Alltech Associates, Deerfield, Ill.).

### B. Reagents

Acetonitrile, HPLC grade, (J. T. Baker, Phillipsburg, N.J.), phosphoric acid, certified A.C.S., (Fisher Scientific, Fair Lawn, N.J.), and sodium hydroxide, AR grade, (Mallinckrodt Inc., St. Louis). Mobile phase: 30% $CH_3CN$ in 0.1% $H_3PO_4$ and adjust pH to 4 with 10% NaOH for triamterene and 10% $CH_3CN$ in 0.1% $H_3PO_4$ for hydroxytriamterene sulfate. The solvent is filtered through a Whatman® #2 paper filter (VWR Scientific Inc., San Francisco) and degassed under vacuum.

### C. Standards

Triamterene can be obtained from Smith Kline and French Laboratories, Philadelphia, Pa., hydroxytriamterene sulfate was a gift from Rohm Pharma, Darmstadt, West Germany; and hydroflumethiazide can be obtained from Bristol Laboratories, Syracuse, N.Y.

The stock standard is prepared as follows: 3.15 mg of triamterene is dissolved in 100 mℓ of methanol and 7.62 mg of hydroxytriamterene is dissolved in 100 mℓ of dimethylformamide. The solution is stable at 4°C for at least 6 months. The stock standard is used for preparing urine standards. A 500 times dilution of triamterene and 200 times dilution of hydroxytriamsulfate is used as the plasma stock standard.

A fortified volume of 0, 1, 2, 4, 8, and 15 µℓ of triamterene stock and 0, 2, 5, 10, 20, 40, 80, and 150 µℓ of hydroxytriamterene sulfate stock is added to 200 µℓ urine blanks in culture tubes to yield 0, 0.16, 0.32, 1.26 and 2.36 µg/mℓ of triamterene and 0, 0.76, 1.91, 3.81, 7.69, 15.24, 30.48, and 57.15 µg/mℓ of hydroxytriamterene sulfate.

A stock internal standard is made by dissolving 30 mg hydroflumethiazide in 100 mℓ of CH₃CN. The working internal standard for urine and plasma is diluted with acetonitrile 100 and 1000 times, respectively.

Quality control specimens are prepared in-house in the same manner as the plasma or urine standard. Proper low, medium, and high concentration specimens are prepared and stored at $-30°C$ until assay.

### D. Procedure

To 200 µℓ of plasma sample, plasma standard, control standard, or control specimen, add 400 µℓ of acetonitrile working internal standard. Vortex mix the mixture for 10 sec and then centrifuge for 10 min. Inject 20 µℓ of the supernatant onto the column and elute with mobile phase at a flow rate of 2.0 mℓ/min.

Urine sample: pipette 200 µℓ of urine sample into a culture tube, add 200 µℓ of internal standard stock solution to each tube. Vortex and inject 10 µℓ of the resulting sample onto the column through the WISP® tray. For triamterene, the fluorescence detector is set at a wavelength of 365 nm for excitation and 440 nm for emission. For hydroxytriamterene sulfate, excitation and emission wavelength are set at 350 and 440 nm, respectively. Detector sensitivity is as follows: photomultiplier (PM) gain 2 and sensitivity 3 for triamterene, and PM gain 2 and sensitivity 10 for hydroxytriamterene sulfate for urine. A 20-fold sensitivity is set for both triamterene and hydroxytriamterene sulfate in plasma. Triamterene and internal standard have retention times of 6.0 and 4.0 min, respectively (Figures 1 and 2). The retention time of hydroxytriamterene sulfate and internal standard are 10 and 14 min, respectively.

### E. Calculation

Calibration graphs are constructed from spiked plasma and urine samples using the same procedure described above. The peak height ratios (drug:internal standard) are plotted vs. drug concentration in ng/mℓ (or µg/mℓ), and the calibration graph is used for the calculation of plasma or urine concentration in human subjects.

## IV. RESULTS

### A. Linearity

Peak height ratios of drug to internal standard are linearly related from 0.32 to 31.5 ng/mℓ for plasma triamterene, 1.91 to 114.30 ng/mℓ for plasma hydroxytriamterene sulfate, 0.15 to 2.36 µg/mℓ for urinary triamterene, and 0.76 to 57.15 µg/mℓ for urinary hydroxytriamterene sulfate.

### B. Recovery

Recovery ranges from 97 to 118% for both triamterene and hydroxytriamterene sulfate in plasma.

### C. Interference

No interference has been noted for a number of patients taking triamterene due to the selectivity of the detector.

### D. Precision

The precision for triamterene and hydroxytriamterene sulfate is less than 3% coefficient of variation for both within-day and day-to-day analyses.

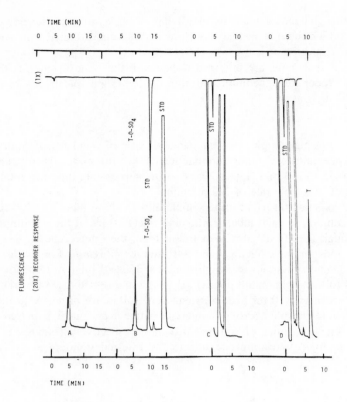

FIGURE 1. Chromatograms of (A) plasma blank, (B) plasma standard containing 10.5 ng/mℓ of hydroxytriamterene sulfate, (C) plasma blank containing internal standard, and (D) plasma blank containing 6.3 ng/mℓ of triamterene.

### E. Metabolites

Triamterene undergoes hydroxylation and subsequent sulfation to hydroxytriamterene sulfate which is the major metabolite. Both parent drug and metabolite can be quantitatively measured in plasma and urine. Hydroxytriamterene can be detected in both solvent systems (assay limit 20 ng/mℓ in plasma and 0.5 µg/mℓ in urine). We could not detect a hydroxytriamterene peak in either plasma or urine from healthy volunteers.

## V. COMMENTS

The therapeutic concentration range for triamterene has not been defined. Oral dosage of 100 mg triamterene results in a peak plasma level of about 125 ng/mℓ. The concentration of the metabolite, hydroxytriamterene sulfate, at multiple time point was about 5 to 10 times higher than that for triamterene. Less than 10% of the drug is excreted unchanged.[5]

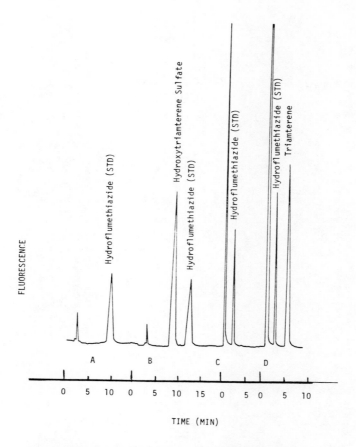

FIGURE 2. Chromatograms of (A) urine blank containing internal standard, (B) urine standard containing 1.91 µg/mℓ of hydroxytriamterene sulfate, (C) urine blank containing internal standard, and (D) urine standard containing 1.26 µg/mℓ of triamterene.

## REFERENCES

1. **Thompson, E. J., Torres, E., Grosberg, S. J., and Maldonado, M. M.,** Effect of triamterene on potassium excretion in cirrhotic patients receiving furosemide, *Clin. Pharmacol. Ther.*, 21, 392, 1976.
2. **Knauf, H., Mohrke, W., Mutschler, E., and Volger, K. D.,** Zur Bioverfugbarkeit von hydrochlorothiazid und triamteren airs fertigarzneimittelin, *Arzheim. Forsch./Drug Res.*, 2, 1001, 1980.
3. **Leilich, G., Knauf, H., Mutschler, E., and Volger, K. D.,** Influence of triamterene and hydroxytriamterene sulfuric acid ester on diuresis and saluresis in rats after oral and intravenous application, *Arzneim. Forsch./Drug Res.*, 30, 949, 1980.
4. **Lin, E. T., Sorgel, F., Hasegawa, J., and Benet, L. Z.,** HPLC analysis of triamterene and metabolites in plasma, *Abstr. Acad. Pharm. Sci.*, 10, 120, 1980.
5. **Hasegawa, J., Lin, E. T., Williams, R. L., Sorgel, F., and Benet, L. Z.,** Pharmacokinetis of triamterene and its metabolite in man, *J. Pharmacokinet. Biopharm.*, 10, 507, 1982.

Chapter 25

# EXOGENOUS GLUCOCORTICOIDS, PREDNISONE, AND PREDNISOLONE IN PLASMA

### Emil T. Lin

## TABLE OF CONTENTS

I.   Introduction ................................................................. 130

II.  Principle .................................................................... 130

III. Materials and Methods ....................................................... 130
     A.   Equipment ............................................................. 130
     B.   Reagents .............................................................. 130
     C.   Standards ............................................................. 130
     D.   Procedure ............................................................. 131
     E.   Calculation ........................................................... 132

IV.  Results ..................................................................... 132
     A.   Linearity ............................................................. 132
     B.   Recovery .............................................................. 132
     C.   Interference .......................................................... 132
     D.   Precision ............................................................. 132
     E.   Temperature ........................................................... 132
     F.   Metabolites ........................................................... 132

V.   Comments .................................................................... 132

References ....................................................................... 133

## I. INTRODUCTION

Among the steroids determined for clinical purposes, the glucocorticoids are the most often requested. The determination of cortisol or its metabolites is indispensable for the elucidation of disease states such as hypocorticism, hypercorticism, and congenital adrenal hyperplasia.[1,2] The cortisol/cortisone ratio in the amniotic fluid has also been identified as an important predictor of respiratory distress syndrome.[3]

The quality and quantity of information obtained from liquid chromatography (LC) is significantly better than that obtained from RIA procedures. In LC, multiple compounds and their metabolites can be assayed in a single run and a chromatogram allows the chromatographer to consider the possibility of interfering substances. Unless the interfering substances coelute exactly with the compounds of interest, peaks will be skewed or fused and thus alert the chromatographer that the result is suspect. With RIA such clues are not available. In addition, the LC peak can be collected for further identification, a process impossible with RIA.

## II. PRINCIPLE

The steroids are extracted from 1 m$\ell$ of plasma with methylene chloride/ether, washed with acid and base, and separated isocratically on a normal-phase silica column with a mobile phase consisting of methylene chloride, methanol, tetrahydrofuran, and acetic acid. The drugs are detected by their absorption at 254 nm and quantitated from peak height ratios.

## III. MATERIALS AND METHODS

### A. Equipment

Liquid chromatography systems equivalent to the following should be used: a Model 6000A pump (Waters Associates, Inc., Milford, Mass.) equipped with an automatic sampler WISP® 710B (Waters Associates), an LC-15 UV-Detector (Perkin-Elmer Corp., Norwalk, Conn.), and a dual pen recorder (Tegal Scientific, Inc., Martinez, Calif.). The 3.2 × 250 mm column is packed with 5 μ silica (Altex Scientific, Berkeley, Calif.).

### B. Reagents

Methylene chloride, tetrahydrofuran and methanol, HPLC grade (Fisher Scientific Co., Tustin, Calif.), glacial acetic acid, A. C. S. grade (J. T. Baker, Phillipsburg, N.J.), and anhydrous ethyl ether (Mallinckrodt, Inc., St. Louis) are used. Mobile phase: 1.5% methanol, 1.0% tetrahydrofuran, and 0.5% glacial acetic acid in methylene chloride. The solvent is filtered through Whatman® #2 filter paper (VWR Scientific, San Francisco) and degassed under vacuum.

### C. Standards

Glucocorticoids used as standards: prednisone, prednisolone, cortisone, cortisol, and dexamethasone can be obtained from Sigma Chemical Co., St. Louis.

Stock standard is prepared as follows: 100 mg each of cortisone, cortisol, prednisone, and prednisolone are dissolved in 1 $\ell$ of methanol.

Reference standard: dilute 0.1, 0.2, 0.5, 1, 5, 50, and 100 m$\ell$ of the corticosteroid stock solution to 100 m$\ell$ with methanol.

Add 100 μ$\ell$ of each reference standard to 1 m$\ell$ of plasma to achieve concentrations of 10, 20, 50, 100, 500, 1000, 5000, and 10,000 ng/m$\ell$, respectively.

A stock internal standard is made by dissolving 100 mg of dexamethasone in 1 $\ell$ of

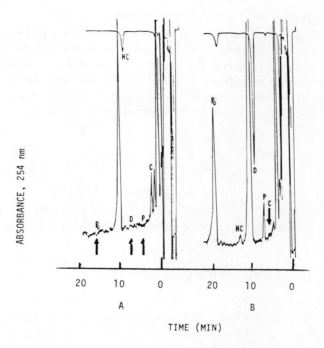

FIGURE 1. Chromatograms of (A) blank human plasma extract containing cortisol (HC) 204 ng/mℓ and cortisone (C) 17 ng/mℓ and (B) plasma sample from the same subject, prednisone treated, containing prednisolone (Po) 265 ng/mℓ, prednisone (P) 25.7 ng/mℓ, and cortisol (HC) < 10 ng/mℓ.

methanol. A tenfold dilution of the stock with methanol is used as the working internal standard.

Quality control specimens are prepared in-house in the same manner as the plasma standards where low, medium, and high concentrations are prepared and stored at −30°C until assay.

### D. Procedure

With a mechanical shaker, extract 1 mℓ of plasma containing 75 µℓ of the internal standard into 9 mℓ of a 2:1 (by volume) mixture of methylene chloride and ether for 20 min. Centrifuge for 10 min and transfer the organic phase into a 16 × 125 mm glass tube. Add 2 mℓ of 0.1 N HCl, shake, centrifuge for 10 min, and remove the aqueous layer. Repeat the same procedure twice more with 1 mℓ 0.1 N NaOH and distilled water. Remove the organic layer to a clean glass tube and evaporate under a stream of nitrogen gas. Dissolve the residue in 200 mℓ of the mobile phase, inject approximately 70 µℓ of the sample onto the column, and elute with the mobile phase at a flow rate of 1.3 mℓ/min. The column temperature is maintained at room temperature and column effluent is monitored at 254 nm. Detector sensitivity is set to 0.004 A full scale. The output of the dual pen recorder is set at 10 and 50 mV. The retention times for cortisone, prednisone, dexamethasone, cortisol, and prednisolone are 6.3, 7.1, 11.7, 14.1, and 20.7 min, respectively (Figure 1).

To prepare a standard curve, add the steroids to normal human plasma and correct for endogenous cortisol and cortisone present in the plasma. This method is preferred to the time-consuming removal of endogenous glucocorticoids which is often incomplete.

## E. Calculation

Plasma standards are carried throughout the procedure and used to prepare a standard curve based on the peak height ratio of steroids to the internal standard dexamethasone.

Because the standard curve is prepared over a wide concentration range, calculate least squares regressions for the logarithm peak height ratio vs. the logarithm of steroid concentration. This serves to increase the weight of the lower concentration values. If no weighting scheme is used, these lower values would be insignificant in calculating the best fits for the standard curve.

## IV. RESULTS

### A. Linearity

For routine analysis of 1 ml of plasma, 10 to 10,000 µg/l of the four compounds can be measured. The standard curve for these compounds shows good linearity ($r_2 = 0.998$).[4]

### B. Recovery

Analytical recovery for each compound exceeds 75%.

### C. Interference

Potential interference by 20 other steroids and 26 commonly used drugs were investigated by chromatographing each compound and extracting those with retention times similar to one of the five steroids studied. Methylprednisolone has a retention time slightly less than prednisolone and can be measured by our assay. Aldosterone and 11-keto-etiocholanolone interfere with dexamethasone, but they are present and extracted in such low concentrations, (<0.5 µg/l) that they should negligibly affect the accuracy of this method. Theophylline has a retention time similar to dexamethasone and would interfere with corticosteroid determination by giving a falsely high peak for the internal standard.

### D. Precision

Coefficients of variation for both intra- and inter-day analyses are less than 11% for the four steroids.[4]

### E. Temperature

The use of elevated temperature is not recommended, since the solvent system mainly consists of highly volatile methylene chloride.

### F. Metabolites

Prednisolone and prednisone, and cortisone and cortisol are interconvertible in the body. Regardless of the compound administered, both the drug and its metabolite can be separated by this procedure.

## V. COMMENTS

Recently, we tested a simplified extraction method.[5] A plasma sample containing internal standard is mixed with 1 ml of 0.1 N sodium hydroxide and poured into a Clin Elut Extube® (#1003, Analytichem International, Harbor City, Calif.) and allowed to be adsorbed for 2 to 3 min. Methylene chloride (2 × 5 ml portions) is then used to elute the drugs and internal standard from the column. After evaporation to dryness under nitrogen gas, the residue is dissolved in 200 µl of the mobile phase and one third to one half of the sample is injected onto the column. By using this extraction method, a good correlation, with coefficients of variation of less than 14%, is obtained for all four corticosteroids. This extraction method greatly reduces the extraction time.

## REFERENCES

1. **Eldy, R. L., Jones, A. L., Gilliand, P. F., Ibarra, J. D., Jr., Thompson, J. Q., and McMurry, J. F., Jr.,** Cushing's syndrome: a prospective study of diagnostic methods, *Am. J. Med.*, 55, 621, 1973.
2. **Finkelstein, M. and Schaefer, J. M.,** Inborn errors of steroid biosynthesis, *Physiol. Rev.*, 59, 353, 1979.
3. **Smith, B. T., Worthington, D., and Maloney, A. H. A.,** Fetal lung maturation. III. The amniotic fluid cortisol/cortisone ration in preterm human delivery and the risk of respiratory distress syndrome, *Obstet. Gynecol.*, 49, 527, 1977.
4. **Frey, F. J., Frey, B. M., and Benet, L. Z.,** Liquid chromatographic measurement of endogenous and exogenous glucocorticoids in plasma, *Clin. Chem.*, 25, 1944, 1979.
5. **Stewart, J. T., Honigberg, I. L., Turner, B. M., and Davenport, D. A.,** Improved sample extraction before liquid chromatography of prednisone and prednisolone in human serum, *Clin. Chem.*, 28, 2326, 1982.

Chapter 26

# 5-FLUOROCYTOSINE BY ULTRAVIOLET DETECTION

## George R. Gotelli and Jeffrey H. Wall

## TABLE OF CONTENTS

| | | |
|---|---|---|
| I. | Introduction | 136 |
| II. | Principle | 136 |
| III. | Materials and Methods | 136 |
| | A. Equipment | 136 |
| | B. Reagents | 136 |
| | C. Standards | 136 |
| | D. Procedure | 136 |
| | E. Calculations | 136 |
| IV. | Results | 137 |
| | A. Optimization of Chromatography | 137 |
| | B. Linearity | 138 |
| | C. Recovery | 138 |
| | D. Reproducibility | 138 |
| | E. Interference | 138 |
| V. | Comments | 138 |
| References | | 138 |

## I. INTRODUCTION

The fluorinated pyrimidine, 5-fluorocytosine, is used in the treatment of fungal infections. The drug is not significantly metabolized in man and up to 95% of the parent compound is excreted unchanged in the urine. The toxic effects of this drug, such as bone marrow depression and anemia, as well as the need to insure an adequate plasma concentration, require that a suitable assay method be available. Microbiological assay methods are tedious[1] and gas-liquid chromatographic techniques are lengthy and require derivitization.[2] Liquid chromatography (LC) does not require derivitization and the samples can be prepared rapidly (Figure 1). This method is a modification of the method of Miners et al.[3]

## II. PRINCIPLE

Serum proteins are precipitated with trichloracetic acid containing the internal standard 5-methyl cytosine. The supernatant is chromatographed directly on a reversed-phase column. The drugs are detected at 276 nm and quantitated by peak heights ratios.

## III. MATERIALS AND METHOD

### A. Equipment

A series 1 pump (Perkin-Elmer Corp., Norwalk, Conn.), a Rheodyne® 7105 valve (Rheodyne, Cotati, Calif.), a Perkin-Elmer 123 recorder, a temperature controlled oven (Perkin-Elmer, LC-100), maintained at 30°C, a 25 cm reversed-phase octadecylsillane column similar to a Waters® C18 μBondapak® (Waters Associates Inc., Milford, Mass.), and a detector similar to a Perkin-Elmer Model LC-55 set at 276 nm are used. The column is eluted with 10 mm/ℓ phosphate buffer (pH 7.0) at a flow rate of 1.5 mℓ/min and column effluent detected at 276 nm.

### B. Reagents

The mobile phase consists of 10 mmol/mℓ dipotassium hydrogen phosphate. The pH is adjusted to 7.0 with concentrated phosphoric acid.

### C. Standards

The 5-fluorocytosine was obtained from Hoffman-LaRoche Inc., Nutley, N.J. The internal standard 5-methylcytosine was purchased from Sigma Chemical Co., St. Louis. Prepare a 50 μg/mℓ solution in 100 g/mℓ trichloroacetic acid. A reference standard consists of 25 μg/mℓ each of 5-fluorocytosine and 5-methylcytosine in 100 g/mℓ trichloroacetic acid.

### D. Procedure

Add 0.5 mℓ of unknown serum and 0.5 mℓ of internal standard solution to a tube and vortex well. Centrifuge briefly to separate the supernatant, pour the supernatant into a clean tube and add 0.5 mℓ of ethyl acetate. Vortex for 1 min and centrifuge briefly. Aspirate to waste the upper ethyl acetate layer and inject 20 μℓ of the trichloroacetic acid layer onto the column. Following elution of the unknown inject 20 μℓ of the reference standard.

### E. Calculations

Calculate a response factor (RF) from the reference standard chromatogram as follows:

$$\frac{\text{peak height of Int. Std.}}{\text{peak height of 5-fluorocytosine}} = \text{R.F.}$$

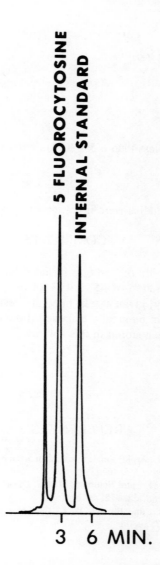

FIGURE 1. Chromatogram of a blood sample containing 36 μg of 5-fluorocytosine per milliliter (obtained in the authors' laboratory).

Calculate the unknowns from the respective chromatogram as follows:

$$\frac{\text{peak height of unknown 5-fluorocytosine}}{\text{peak height of Int. Std.}} \times RF \times 50 = \text{unknown 5-fluorocytosine in } \mu g/m\ell$$

## IV. RESULTS

### A. Optimization of Chromatography

This procedure requires the use of "uncapped" C-18 reversed-phase columns such as Waters® or Whatman® brands. "Capped" columns, with no exposed polar silica sites, will not retain 5-fluorocytosine. The ethyl acetate extraction step removes many blood constituents which would elute from the column after the 5-fluorocytosine and internal standard peaks.

## B. Linearity
This method is linear to 200 µg/mℓ.

## C. Recovery
Analytical recovery is >98%.

## D. Reproducibility
The day-to-day precision is less than 6.5% (CV). Within-day precision is less than 4.0% (CV).

## E. Interference
Interference and accuracy studies were investigated by Miners et al.[3]

## V. COMMENTS

5-Fluorocytosine has a half-life of 3 to 6 hr in individuals with normal renal function. It is found in CSF at a concentration of 65 to 95% of the plasma concentration. The usual dosage of 50 to 150 mg/kg/day, given at 6 hr intervals, results in peak concentrations of 50 to 100 µg/mℓ. Column life can be greatly extended if the column is washed briefly with a 50% methanol solution and then stored in this solution.

## REFERENCES

1. **Kaspar, R. L. and Drutz, D. J.**, Rapid, simple bioassay for 5-fluorocytosine in the presence of amphotericin B, *Antimicrob. Agents. Chemather.*, 7, 462, 1975.
2. **Harding, S. A., Johnson, G. F., and Solomon, H. M.**, Gas-chromatographic determination of 5-fluorocytosine in serum, *Clin. Chem.*, 22, 772, 1976.
3. **Miners, J. D., Foenander, T., and Birkett, D. J.**, Liquid-chromatographic determination of 5-fluorocytosine, *Clin. Chem.*, 26, 117, 1980.

Chapter 27

# SIMULTANEOUS DETERMINATION OF METHOTREXATE AND ITS METABOLITES IN PLASMA, SALIVA, AND URINE

**Mei-Ling Chen and Win L. Chiou**

## TABLE OF CONTENTS

I. Introduction .................................................................. 140

II. Principle ..................................................................... 140

III. Materials and Methods ...................................................... 140
    A. Equipment ............................................................. 140
    B. Reagents .............................................................. 141
    C. Standards ............................................................. 141
    D. Procedure ............................................................. 141
    E. Calculation ........................................................... 141

IV. Results ...................................................................... 141
    A. Optimization of Chromatography ....................................... 142
    B. Linearity ............................................................. 143
    C. Recovery .............................................................. 143
    D. Reproducibility ....................................................... 144
    E. Interference .......................................................... 144

V. Comments .................................................................... 144

Acknowledgment ................................................................. 145

References ..................................................................... 145

## I. INTRODUCTION

Methotrexate (MTX; 4-amino-$N^{10}$-methylpteroylglutamic acid), a potent antifolate, is widely used for the treatment of various malignant diseases as well as nonneoplastic disorders.[1] With the advent of high-dose therapy followed by leucovorin rescue, plasma monitoring of MTX levels becomes extremely important to allow early detection of patients at high risk of toxicity.[1-4]

A number of assay techniques are available; they include fluorometry,[5] competitive protein binding,[6-8] enzyme inhibition assay,[9,10] radioimmunoassay,[11-13] radioassay,[14] enzyme immunoassay,[15,16] and high-performance liquid chromatography (HPLC).[17-29] However, it has been reported[26,27] that most non-HPLC methods lack specificity due to the potential interference of active metabolites, such as 7-hydroxymethotrexate (7-OH-MTX) and 4-amino-4-deoxy-$N^{10}$-methylpteroic acid (APA). Furthermore, these non-HPLC assays are not capable of quantitating the two major metabolites in biological samples.

As for published HPLC methods, it appears that they have one or more of the following limitations. For instance, 1 m$\ell$[17,19,22,24] to 3 m$\ell$[20,21] of plasma or serum is usually needed. The sample preparations involving extraction, evaporation, and reconstitution are relatively complex and may take more than 20 min[19,22,23] prior to chromatography. Applicability for the determination of 7-OH-MTX[17] and APA[19,20,22,25] in plasma was not often studied. In addition to the relatively low sensitivity for APA, its retention times were always as long as 20 to 30 min.[21,23,24] Similar problems also occurred to 7-OH-MTX in some cases.[29] Only a few HPLC methods reported their application for urine samples.[18,21,24] None of the published methods have shown their feasibility for saliva analysis. Thus, a simple, sensitive, and micro-HPLC assay is described here for the simultaneous determination of MTX, 7-OH-MTX, and APA in plasma, saliva, and urine.

## II. PRINCIPLE

Plasma, serum, or saliva samples were deproteinized with acetonitrile. After centrifugation, the supernatant is extracted with isoamyl alcohol and ethyl acetate. An aliquot of the resultant aqueous solution is injected onto a cation-exchange column and the drug and its metabolites are eluted with an acetonitrile-phosphate buffer mobile phase. The effluent is monitored by a UV-detector set at 313 nm. For urine and plasma samples containing higher concentrations of MTX and its metabolites, only simple deproteinization is needed prior to the chromatography.

## III. MATERIALS AND METHODS

### A. Equipment

The liquid chromatographic system consists of a solvent delivery pump (Model M6000A, from Waters Associates, Milford, Mass. or Model 110A, from Beckman Instruments, Inc, Berkeley, Calif.), a fixed-wavelength detector with 313-nm filter (Model 440 from Waters Associates or Model 160 from Beckman Instruments), a syringe loading sample injector (Model 7125, Rheodyne, Cotati, Calif.), and an ion-exchange column (Partisil® PXS 10/25 SCX, 25 cm × 4.6 mm i.d., particle size 10 µg, available from Whatman, Inc., Clifton, N.J.). The output from the detector is connected to a 10-mV potentiometric 25.4-cm recorder (Linear Instruments, Irvine, Calif.).

The mobile phase is prepared by mixing 10 parts of acetonitrile with 90 parts of 0.02 $M$ monobasic ammonium phosphate solution acidified with phosphoric acid (0.2%). This is pumped through the HPLC system at a flow-rate of 2 m$\ell$/min. The recorder chart speed is

10 cm/hr.[30] All experiments are carried out at ambient temperature. The optimal composition of mobile phase may vary with the column used. For example, apart from 10% acetonitrile, a higher concentration of ammonium phosphate (0.035 $M$ together with a flow-rate of 1 m$\ell$/min is found to be most satisfactory for a different column from the same manufacturer.

### B. Reagents

All reagents are of analytical grade. Ammonium phosphate, phosphoric acid, and glass-distilled acetonitrile can be purchased from Fisher Scientific (Fair Lawn, N.J.). Isoamyl alcohol is from J. T. Baker (Phillipsburg, N.J.), and ethyl acetate from Burdick and Jackson Laboratories (Muskegon, Mich.). Most drugs tested for potential interferences of the assay are obtained from the Hospital Pharmacy, University of Illinois (Chicago).

### C. Standards

MTX and APA are supplied by Dr. Ven L. Narayanan from the National Institute of Health (Bethesda, Md.) and Dr. Maharaj K. Raina from the Lederle Laboratories (Pearl River, N.Y.). The purified 7-OH-MTX is supplied by Dr. David Johns from the National Cancer Institute and Dr. Kenneth K. Chan from the University of Southern California (Los Angeles). Additional samples used for routine standard curve study can be isolated according to the procedure of Watson et al.[19]

Standard solutions (1 $\mu$g/m$\ell$ to 10 mg/m$\ell$) of MTX and APA are prepared in distilled water. 7-OH-MTX purified from DEAE-cellulose column and dissolved in 10 m$M$ Tris-HCl buffer (pH 7.5) is used for spiking directly. Its concentration is determined by comparing the HPLC peak height with those from authentic samples. All standard solutions should be stored at 4°C.

### D. Procedure

Plasma, serum, or saliva (0.2 m$\ell$) from normal subjects or patients is pipetted into 13 × 100 mm screw-capped culture tubes. The deproteinization is carried out by adding 0.5 m$\ell$ (0.4 m$\ell$ is found to be sufficient later) of acetonitrile, followed by vortexing for 10 sec and centrifugation at 800 g for 2 min. The entire supernatant is poured into a glass tube which has a tapered base. After addition of 100 $\mu\ell$ of isoamyl alcohol and 1 m$\ell$ of ethyl acetate, the tube is vortexed for 10 sec and then centrifuged at 800 g for 4 min. About 10 to 30 $\mu\ell$ from the lower aqueous portion is injected onto the column.

Urine samples are prepared by the same deproteinization procedures as described above. Since concentrations in urine are usually much higher, the deproteinized supernatant (20 to 50 $\mu\ell$ can be injected directly onto the column. The one-step deproteinization method can also be used for plasma samples with higher concentrations of drug and the metabolites. In this case, plasma or serum (0.05 to 0.1 m$\ell$ can be deproteinized with 2 vol of acetonitrile and the supernatant (50 $\mu\ell$) analyzed directly.

### E. Calculation

The use of a micrometer (Vernier Caliper from Fisher Scientific, Chicago) has been shown to increase the accuracy of peak height measurements.[31] Standard curves are constructed by supplementing blank human plasma, saliva, and urine with known concentrations of MTX, 7-OH-MTX, and APA. Determination of the concentration of unknown MTX and its two matabolites in biological samples is made by direct comparison with the data obtained from the standards.

## IV. RESULTS

Chromatograms from blank human plasma, saliva, urine, as well as those spiked with

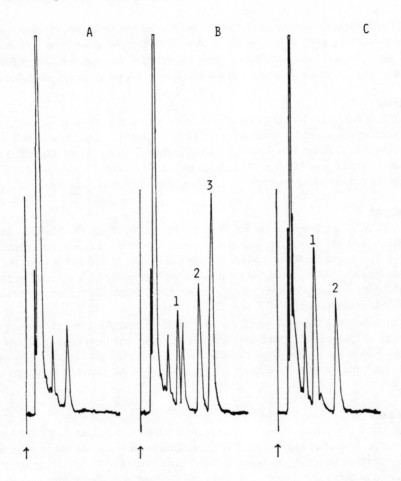

FIGURE 1. Chromatograms of extracts from (A) blank human plasma; (B) plasma spiked with 0.5 μg/mℓ of MTX, 7-OH-MTX, and APA; (C) patient plasma collected at 8 hr after the end of i.v. infusion for 25 hr on a dose of 750 mg/m² MTX. Peaks 1 = 7-OH-MTX, 2 = MTX, and 3 = APA. The arrow marks the point of injection. Detector sensitivity was 0.005 A.U.F.S. (absorbance unit full scale) and recorder chart speed was 20 cm/hr.

known concentrations of MTX, 7-OH-MTX, and APA, together with plasma from a patient on MTX therapy are shown in Figures 1 and 2. The peak shape from MTX, 7-OH-MTX, or APA is all symmetrical with no interferences from endogenous substances. Although there is an endogenous peak between 7-OH-MTX and MTX, it does not affect the present assay. The retention times for MTX, 7-OH-MTX, and APA are 7, 5, and 9 min, respectively.

A higher sensitivity is obtained with salivary samples than plasma as shown in Figure 2. This might be in part due to the lower content of electrolytes in saliva which reduced the final aqueous volume after extraction.

## A. Optimization of Chromatography

Under the above HPLC condition, the detection limits with a 20-μℓ injection volume for MTX, 7-OH-MTX, and APA in plasma are approximately 15, 25, and 10 ng/mℓ, respectively. Higher sensitivity can be obtained when 0.5 mℓ rather than 1 mℓ of ethyl acetate is used for extraction. This is primarily attributed to the smaller volume of the final aqueous

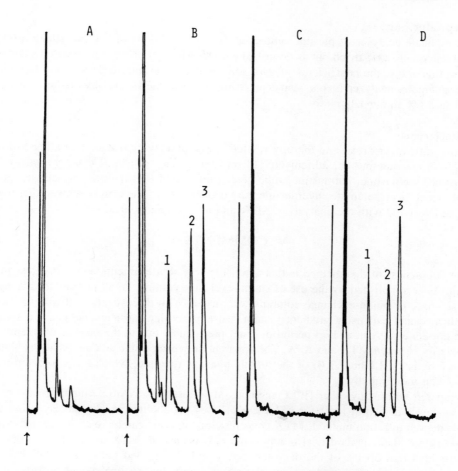

FIGURE 2. Chromatograms from (A) blank human saliva extract; (B) extract of saliva spiked with 0.5 μg/mℓ of MTX, 7-OH-MTX, and APA; (C) deproteinized blank human urine; (D) deproteinized urine spiked with 5 μg/mℓ of MTX, 7-OH-MTX, and APA. The arrow marks the point of injection. Detector sensitivity setting was 0.005 A.U.F.S. and recorder chart speed was 20 cm/hr.

solution after extraction. The above modification could result in detection limits down to 10 ng/mℓ for MTX, 20 ng/mℓ for 7-OH-MTX, and 5 ng/mℓ for APA. With the recent introduction of Beckman® fixed-wavelength detector (Model 160), it is found that the signal-noise ratio for the three compounds can be further enhanced by using its digital filter in the quantitation.

### B. Linearity

Standard curves are linear over the concentration range (0.1 to 10 μg/mℓ) studied for the three compounds using the extraction procedure. As to deproteinized plasma samples, linearity has been established down to 0.05 μg/mℓ for MTX.

### C. Recovery

The extraction efficiencies for plasma samples are 70, 50, and 77% for MTX, 7-OH-MTX, and APA, respectively. Recoveries for saliva samples are higher, being 98, 61, and 79% for these three compounds, respectively. The recovery from the simple one-step deproteinization method is virtually 100%.

## D. Reproducibility

Six replicate analyses of plasma samples at the concentrations of 0.1 and 10 µg/mℓ for three compounds are carried out as described earlier. Although there is no internal standard used in this assay, the coefficients of variation for inter- and intra-day assay are less than 8%. For samples analyzed by one-step deproteinization, the coefficient of variation is usually less than 3.5% in our laboratory.

## E. Interference

Many anticancer drugs and therapy-related compounds were tested, including 5-fluorouracil, 6-mercaptopurine, adriamycin, bleomycin sulfate, cisplatin, cyclophosphamide, vincristine, vinblastine, carmustine, folic acid, folinic acid (leucovorin), 5-methyltetrahydrofolic acid, acetazolamide, hydralazine, and trimethoprim. The results showed that none of them interfered with the analysis of MTX and its two metabolites.

## V. COMMENTS

The uniqueness of this method is that a considerably shorter retention of APA (less than 10 min) is accomplished by the use of cation-exchange column. In all the previous assays, reverse-phase or anion-exchange columns were used. With the exception of using a more complicated and expensive gradient system,[29] it appears that most reverse-phase columns could not elute APA in a short period of time, presumably due to the marked difference in the polarity between MTX and APA. The apparent drawback of the anion-exchange column is that a higher pH (about 7.0) of the mobile phase is required, which tends to deteriorate the column more rapidly.

Compared with many other HPLC methods, the present assay offers higher recovery as well as lower detection limit for MTX and its two metabolites in plasma. By using the simple deproteinization method, MTX concentrations in urine can be readily detected down to 0.1 µg/mℓ. This method is also applicable to plasma with MTX levels above 0.05 µg/mℓ provided that 50 µℓ of the deproteinized sample is injected onto the HPLC system. Therefore, analysis of drug levels in plasma or serum by the one-step deproteinization procedure might be adequate for routine monitoring in high-dose therapy.[1]

Acetonitrile appears to be an ideal deproteinizing agent. This volume ratio of 2.5 between acetonitrile and sample is satisfactory to assure virtual completeness of the deproteinization process. Such a simple deproteinization method has been successfully used in the assay of creatinine,[32] gentamicin,[33] procainamide,[34] tolbutamide,[35] furosemide,[36] and other drugs developed earlier in this laboratory. If a higher sensitivity is desired, however, a volume ratio of 2 between acetonitrile and sample is also sufficient for deproteinization purpose.

In an attempt to increase the sensitivity of the assay, efforts have been made by acidification or alkalinization during the extraction procedure. Surprisingly, the peak heights were all decreased, indicating the reduced extraction efficiency in both cases. Addition of isoamyl alcohol was found to enhance the sensitivity by 1.5- to 2-fold for the three compounds. In view of the high viscosity of this reagent, it is recommended that appropriate amounts of isoamyl alcohol be premixed with ethyl acetate before extraction. Careful sample preparation prior to injection onto the column is essential in reducing analytical errors. Clinical observations[37] have revealed that some patients do not produce APA metabolite; in this case, APA can be employed as an internal standard for the assay.

The method described here permits a rapid, simultaneous determination of MTX and its two active metabolites in biological fluids. It can also be applied to biological samples from animals, such as dogs, rabbits, and rats. The sample preparation prior to chromatography is simple and no evaporation or reconstitution step is needed. In view of its simplicity, specificity, and sensitivity, the method is of use in pharmacokinetic studies and is suitable for routine monitoring of serum levels of MTX as well as its two metabolites, 7-OH-MTX, and APA.

## ACKNOWLEDGMENT

The preparation of this chapter was in part supported by a grant from the National Cancer Institute of the NIH, CA/GM 29754-02.

## REFERENCES

1. **Bleyer, W. A.**, The clinical pharmacology of methotrexate, *Cancer*, 41, 36, 1978.
2. **Nirenberg, A., Masende, C., Mehta, B. M., Gisolfi, A. L., and Rosen, G.**, High-dose methotrexate with citrovorum factor rescue: predictive value of serum methotrexate concentrations and corrective measures to avert toxicity, *Cancer Treat. Rep.*, 61, 779, 1977.
3. **Stoller, R. G., Hard, K. R., Jacobs, S. A., Rosenberg, S. A., and Chabner, B. A.**, Use of plasma pharmacokinetics to predict and prevent methotrexate toxicity, *N. Engl. J. Med.*, 297, 630, 1977.
4. **Frei, E., III, Jaffe, N., Tattersall, M. H. N., Pitman, S., and Parker, L.**, New approaches to cancer chemotherapy with methotrexate, *N. Engl. J. Med.*, 292, 846, 1975.
5. **Chakrafarti, S. G. and Bernstein, I. A.**, A simplified fluorometric method for determination of plasma methotrexate, *Clin. Chem.*, 15, 1157, 1969.
6. **Myers, C. E., Lippman, M. E., Eliot, H. M., and Chabner, B. A.**, Competitive protein binding assay for methotrexate, *Proc. Natl. Acad. Sci. U.S.A.*, 72, 3683, 1975.
7. **Kamen, B. A., Takach, P. L., Vaten, R., and Caston, J. D.**, Radiochemical-ligand binding assay for methotrexate, *Anal. Biochem.*, 70, 54, 1976.
8. **Erlichman, C., Donehower, R. C., and Myers, C. E.**, Competitive protein binding assay of methotrexate, *Methods Enzymol.*, 84, 447, 1982.
9. **Falk, L. C., Clark, S. M., Kahlman, S. M., and Long, T. F.**, Enzymatic assay for methotrexate in serum and cerebrospinal fluid, *Clin. Chem.*, 22, 785, 1976.
10. **Pesce, M. A. and Bodourian, S. H.**, Enzyme immunoassay and enzyme inhibition assay of methotrexate, with use of the centrifugal analyzer, *Clin. Chem.*, 27, 380, 1981.
11. **Paxton, W. and Rowell, F. J.**, A rapid sensitive and specific radioimmunoassay for methotrexate, *Clin. Chim. Acta*, 80, 563, 1977.
12. **Kamel, R., Landon, J., and Forrest, G. C.**, A fully automated, continuous-flow radioimmunoassay for methotrexate, *Clin. Chem.*, 26, 97, 1980.
13. **Langone, J. J.**, Radioimmunoassay of methotrexate, leucovorin, and 5-methyltetrahydrofolate, *Methods Enzymol.*, 84, 409, 1982.
14. **Hayman, R. J., Fong, H., and Vancer, M. D.**, A simple radiometric assay for methotrexate and other folate antagonists, *J. Lab. Clin. Med.*, 93, 480, 1979.
15. **Al-Bassam, M. N., O'Sullivan, M. J., Bridges, J. W., and Marks, V.**, Improved double-antibody enzyme immunoassay for methotrexate, *Clin. Chem.*, 25, 1448, 1979.
16. **Finley, P. R., Williams, R. J., Griffith, F., and Lichti, D. A.**, Adaptation of the enzyme-multiplied immunoassay for methotrexate to the centrifugal analyzer, *Clin. Chem.*, 26, 341, 1980.
17. **Nelson, J. A., Harris, B. A., Decker, W. J., and Farquhar, D.**, Analysis of methotrexate in human plasma by high-performance liquid chromatography with fluorescence detection, *Cancer Res.*, 37, 3970, 1977.
18. **Wisnicki, J. L., Tong, W. P., and Ludlum, D. B.**, Analysis of methotrexate and 7-hydroxymethotrexate, by high-pressure liquid chromatography, *Cancer Treat. Rep.*, 62, 529, 1978.
19. **Watson, E., Cohen, J. L., and Chan, K. K.**, High-pressure liquid chromatographic determination of methotrexate and its major metabolite, 7-hydroxymethotrexate, in human plasma, *Cancer Treat. Rep.*, 62, 381, 1978.
20. **Lankelma, J. and Poppe, H.**, Determination of methotrexate in plasma by on-column concentration and ion-exchange chromatography, *J. Chromatogr.*, 149, 587, 1978.
21. **Donehower, R. C., Hande, K. R., Drake, J. C., and Chabner, B. A.**, Presence of 2,4-diamino-$N^{10}$-methylpteroic acid after high-dose methotrexate, *Clin. Pharmacol. Ther.*, 26, 63, 1979.

22. **Cohen, J. L., Hisayasu, G. H., Barrientos, A. R., Nayar, M. S. B., and Chan, K. K.,** Reversed-phase high-performance liquid chromatographic analysis of methotrexate and 7-hydroxymethotrexate in serum, *J. Chromatogr.,* 181, 478, 1980.
23. **Canfell, C. and Sadee, W.,** Methotrexate and 7-hydroxymethotrexate: serum level monitoring by high-performance liquid chromatography, *Cancer Treat. Rep.,* 64, 165, 1980.
24. **Wang, Y. M., Howell, S. K., and Benvenuto, J. A.,** Paired-ion high pressure liquid chromatography of methotrexate and metabolites in biological fluids, *J. Liquid Chromatogr.,* 3, 1071, 1980.
25. **Tong, W., Wisnicki, J. L., Horton, J., and Ludlum, D. B.,** A direct analysis of methotrexate, dichloromethotrexate and their 7-hydroxy metabolites in plasma by high pressure liquid chromatography, *Clin. Chim. Acta,* 107, 67, 1980.
26. **Howell, S. K., Wang, Y., Hosoya, R., and Sutow, W. W.,** Plasma methotrexate as determined by liquid chromatography, enzyme-inhibition assay and radioimmunoassay after high-dose infusion, *Clin. Chem.,* 26, 734, 1980.
27. **Buice, R. G., Evans, W. E., Karas, J., Nicholas, C. A., Sidhu, P., Straughn, A. B., Meyer, M. C., and Crom, W. R.,** Evaluation of enzyme immunoassay, radioassay, and radioimmunoassay of serum methotrexate, as compared with liquid chromatography, *Clin. Chem.,* 26, 1902, 1980.
28. **Breithaupt, H., Kuenzlen, E., and Goebel, G.,** Rapid high-pressure liquid chromatographic determination of methotrexate and its metabolites, 7-hydroxymethotrexate and 2,4-diamino-$N^{10}$-methylpteroic acid in biological fluids, *Anal. Biochem.,* 121, 103, 1982.
29. **Cairnes, D. A. and Evans, W. E.,** High-performance liquid chromatographic assay of methotrexate, 7-hydroxymethotrexate, 4-deoxy-4-amino-$N^{10}$-methylpteroic acid and sulfamethoxazole in serum, urine and cerebrospinal fluid, *J. Chromatogr.,* 231, 103, 1982.
30. **Chiou, W. L.,** Advantages of using slower recorder chart speeds in liquid-chromatographic and gas-chromatographic analyses, *Clin. Chem.,* 25, 197, 1979.
31. **Athanikar, N. K., Peng, G. W., Nation, R. L., Huang, S.-M., and Chiou, W. L.,** Rapid quantitative analysis of chlorpheniramine in plasma, saliva and urine by high-performance liquid chromatography, *J. Chromatogr.,* 162, 367, 1979.
32. **Chiou, W. L., Gadalla, M. A. F., and Peng, G. W.,** Simple, rapid and micro high-pressure liquid chromatographic determination of endogenous "true" creatinine in plasma, serum, and urine, *J. Pharm. Sci.,* 67, 182, 1978.
33. **Nation, R. L., Peng, G. W., Chiou, W. L., and Malow, J.,** Comparison of gentamicin $C_1$ and ($C_{1a}$, $C_2$)-levels in patients, *Eur. J. Clin. Pharmacol.,* 13, 459, 1978.
34. **Gadalla, M. A. F., Peng, G. W., and Chiou, W. L.,** Rapid and micro high-performance liquid chromatographic method for simultaneous determination of procainamide and N-acetylprocainamide in plasma, *J. Pharm. Sci.,* 67, 869, 1978.
35. **Nation, R. L., Peng, G. W., and Chiou, W. L.,** Simple, rapid, and micro high-performance high-pressure chromatographic method for the simultaneous determination of tolbutamide and carboxytolbutamide in plasma, *J. Chromatogr.,* 146, 121, 1978.
36. **Nation, R. L., Peng, G. W., and Chiou, W. L.,** Quantitative analysis of furosemide in micro plasma volumes by high-performance liquid column chromatography, *J. Chromatogr.,* 162, 88, 1979.
37. **Chen, M. L. and Chiou, W. L.,** Pharmacokinetics of methotrexate in patients using a specific HPLC assay, *Int. J. Clin. Pharmacol. Ther. Toxicol.,* in press.

Chapter 28

# ANALYSIS OF MISONIDAZOLE AND DESMETHYLMISONIDAZOLE

**Ellen M. Levin and Victor A. Levin**

## TABLE OF CONTENTS

I. Introduction ................................................................. 148

II. Principle .................................................................... 148

III. Materials and Methods ....................................................... 148
    A. Equipment ............................................................. 148
    B. Reagents .............................................................. 148
    C. Standards ............................................................. 149
    D. Procedure ............................................................. 149
    E. Calculation ........................................................... 149

IV. Results ..................................................................... 149
    A. Linearity ............................................................. 149
    B. Recovery .............................................................. 150
    C. Reproducibility ....................................................... 150
    D. Interferences ......................................................... 150

V. Comments .................................................................... 150

Acknowledgment ................................................................. 151

References ..................................................................... 151

## I. INTRODUCTION

The 2-nitromidazole derivatives, misonidazole and desmethylmisonidazole, are compounds currently being investigated for their ability to sensitize hypoxic mammalian cells to irradiation.[1-4] In addition to their property as hypoxic cell sensitizers, prolonged exposure has been shown to be directly cytotoxic to hypoxic cells at concentrations nontoxic to normal aerated cells.[5] Quantitative correlations between nitromidazole cell concentrations and radiation enhancement,[6] as well as with known peripheral nervous system toxicity[4,5,7] have been demonstrated. These findings make it imperative to monitor drug levels in plasma and urine.

Analytical methods available utilizing UV spectrophotometry[8,9] and polarography[8,10-12] do not differentiate between these drugs and their potentially active nitroimidazole metabolites. A previously reported, gas-liquid chromatography-electron capture assay[8] is sensitive and specific for the parent drug but too lengthy for routine use. Thin-layer chromatography and high-pressure liquid chromatography (HPLC) assays have been developed for 2-nitroimidazoles and for the 5-nitroimidazole, metronidazole.[13-17] The present paper describes a reversed-phase high performance LC method for the assay of the nitroimidazoles in plasma and urine with UV detection at 280 nm. This procedure will separate and is specific for misonidazole, desmethymisonidazole, and fluornitroimidazole, and was utilized to determine pharmacokinetics in patients undergoing radiation therapy in conjunction with these nitroimidazole radiosensitizers.

## II. PRINCIPLE

Plasma samples are spiked with fluornitroimidazole (FMM, internal standard) and extracted with ethyl acetate. After centrifugation, the organic layer is removed and evaporated under a nitrogen stream at ambient temperature. The residue is redissolved in mobile phase and an aliquot injected onto a C18 μBondapak® reversed-phase column. The drugs are eluted with an acetonitrile:sodium acetate buffer mobile phase, detected by their UV absorption at 280 nm, identified by their retention times, and quantitated by the peak area ratio of the drug to internal standard.

## III. MATERIALS AND METHODS

### A. Equipment

A Perkin-Elmer (Perkin-Elmer Corp., Norwalk, Conn.) Series 2 HPLC was equipped with a Chromatronix® Model 220 UV fixed wavelength detector (Spectra-Physics, Mountain View, Calif.), Waters® Intelligent Sampler (WISP® Model 710B (Waters Associates Inc., Milford, Mass.), Alltech 600RP 10 μ C18 μBondapak® 25 cm × 4.6 mm, or Waters® 10 μ C18 μBondapak® 30 cm × 10 mm reversed-phase column. A Model 7302 column inlet filter (2 μ filter element, 3 mm × 3 mm) was placed in line between the injection valve and column. A digital data system (e.g., Sigma® 10 Chromatograph Data Station, Perkin-Elmer Corp.) was utilized to integrate peak areas.

### B. Reagents

Aceonitrile (UV grade), distilled in glass (Burdick and Jackson Laboratories, Inc., Muskegon, Mich.), ethyl acetate, (UV grade, Burdick and Jackson), glacial acetic acid (99.7% conc., Mallinckrodt, St. Louis), sodium acetate (anhydrous, analytical reagent grade, J. T. Baker, Phillipsburg, N.J.).

Mobile phase: 10% solution of acetonitrile in 0.45 $M$ sodium acetate buffer (pH 4.5).

Sodium acetate buffer: add 30 g anhydrous sodium acetate to 2000 mℓ double distilled water containing 30 mℓ of glacial acetic acid. Filter mobile phase with nylon-66 membrane filter, 0.45 μ pore size, 25 mm diameter disc (Rainin Instruments, Woburn, Mass.) and degas.

### C. Standards

Desmethylmisonidazole [1-(2-nitroimidazol-1-yl)-3-hydroxy propran-2-ol,Ro 05-9963], Misonidazole [1-(2-nitroimidazol-1-yl)-3-methoxypropran-2-ol,Ro 07-0582], and fluornitroimidazole [1-(2-nitroimidazol-1-yl)-3-fluoropropran-2-ol,Ro 07-0741] were obtained from the National Cancer Institute (Bethesda, Md.). The molecular weights of misonidazole, desmethylmisonidazole, and fluornitroimidazole, are 201, 187, and 189, respectively. All chemicals were of analytical reagent grade quality.

The misonidazole, desmethylmisonidazole, and fluornitroimidazole stock standards are prepared as follows: dissolve 5 mg of each in 10 mℓ of double distilled water in volumetric flasks. The solutions are stable at 0°C for at least 6 months. Working standards are prepared by diluting the stock standards tenfold with double distilled water. These solutions are stable at 4°C for a month. The nitroimidazoles are photosensitive, therefore, samples should be protected from light while working with them.

The standard curve is prepared by adding 10, 100, and 300 μℓ of each working standard and 100 μℓ of FMM (internal standard) to pooled normal drug free plasma to obtain concentrations of 5, 50, and 150 μg/mℓ, respectively, of drug and 50 μg/mℓ concentration of the internal standard.

Control specimens are prepared in-house by pooling patient plasma of known concentrations.

### D. Procedure

Add 5 mℓ of ethyl acetate to 100 μℓ of plasma or urine (urine needs to be diluted 1:10 to 1:50 depending on time sampled and volume) spiked with the internal standard, FMM. Extract and centrifuge at 2000 rpm for 5 min, remove the organic layer, and evaporate under a nitrogen stream at ambient temperature. Redissolve the residue in 100 μℓ of mobile phase inject 20 μℓ onto the chromatograph which is eluted at a flow rate of 1.5 mℓ/min at an inlet pressure of approximately 1500 psi at ambient temperature. The column effluent is monitored at 280 nm with a sensitivity set at 0.02 AUFS (absorption units full scale). The peak areas of the compounds are determined by the Perkin-Elmer Sigma® 10 Data Station. Total chromatographic time is 10 min (Figure 1).

### E. Calculation

Serum standards are carried through the procedure and are used to prepare a standard curve based on relative peak areas. The peak area of both the compound and internal standard are measured. A calibration curve is constructed by plotting the peak area ratio of either misonidazole or desmethylmisonidazole and that of the internal standard (FMM) vs. the known concentration of each standard. The slope of the calibration curve is determined and the concentration of misonidazole or desmethylmisonidazole in the unknown sample computed using the following relationship:

$$\text{Sample concentration } (\mu g/m\ell) = \frac{\text{slope} \times \text{sample peak area}}{\text{Int. Std. peak area}}$$

## IV. RESULTS

### A. Linearity

Peak area ratios of drug to internal standard were linearly related from 5 to 150 μg/mℓ of each drug.

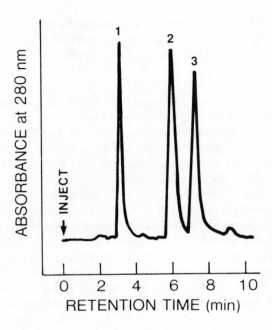

FIGURE 1. High-pressure liquid chromatogram of ethyl acetate extract of plasma. Peaks: (1) desmethylmisonidazole at 3.16 min, (2) fluornitroimidazole at 5.98 min, (3) misonidazole at 7.30 min.

## B. Recovery

The overall extraction recoveries for misonitroimidazole, desmethylmisonidazole, and fluornitroimidazole were 84, 93, and 95%, respectively.

## C. Reproducibility

The precision for misonidazole, desmethylmisonidazole, and fluornitroimidazole equalled 4% coefficient of variation (CV) and 1.6% CV, respectively, for both within day and day-to-day analyses. This analytical method was validated for accuracy by measuring reference calibration standards and three unknown samples containing both drugs in pooled human plasma and urine obtained from the Central Reference Laboratory at the National Cancer Institute. The samples results were reported as $\mu g/m\ell$ and were 96% accurate.

## D. Interferences

Of the many drugs tested, Theodur® theophylline) and Keflex® interfered with fluornitroimidazole. Hemolyzed samples can be assayed without interferences.

## V. COMMENTS

Blood samples obtained should be centrifuged and separated as soon as possible. Urine samples should be collected in clean containers and frozen as soon as possible.

Samples should be kept frozen until used since bacteria growth rapidly destroys misonidazole.

Nitroimidazoles are photosensitive and should be protected from light.

Plastic tubes or containers should not be used since misonidazole may adsorb to different plastics to a different degree.

## ACKNOWLEDGMENT

This study was supported in part by HEW grant CA-13525.

## REFERENCES

1. **Chapman, J. D., Reuvers, A. P., and Borsa, J.,** Effectiveness of nitrofuran derivatives in sensitizing hypoxic mammalian cells to x-rays, *Br. J. Radiol.,* 46, 623, 1973.
2. **Asquith, J. C., Phil, M., Foster, J. L., Wilson, R. L., Ings, R., and McFadzean, J. A.,** Metronidazole ("Flagyl"), a radiosensitizer of hypoxic cells, *Br. J. Radiol.,* 47, 474, 1974.
3. **Dische, S., Saunders, M. L., Flockhart, I. R., Lu, M. E., and Anderson, P.,** Misonidazole — a drug for trial in radiotherapy and oncology, *Int. J. Radiat. Oncol. Biol.,* 5, 851, 1979.
4. **Wasserman, T. H., Phillips, T. L., Johnson, R. J., Gomer, C. J., Lawrence, G. A., Sadee, W., Marques, R. A., Levin, V. A., and Van Raalte, G.,** Initial United States Clinical and Pharmacologic Evaluation of Misonidazole (RO-07-0587), an hypoxic cell radiosensitizer, *Int. J. Radiat. Oncol. Biol.,* 5, 775, 1979.
5. **Brown, J. M.,** Cytotoxic effects of the hypoxic cell radiosensitizers RO-07-0582 to tumor cells in vivo, *Radiat. Res.,* 72, 469, 1977.
6. **Stratford, I. J., Adams, G. E., Horgman, M. R., Sivamany, K., Rajaratnam, S., Smith, E., and Williamson, C.,** The interaction of misonidazole with radiation, chemotherapeutic agents or heat, *Cancer Clin. Trials,* 3, 231, 1980.
7. **Saunders, M. I., Dische, S., Anderson, P., and Flockhart, I. R.,** The neurotoxicity of misonidazole and its relationship to dose, half-life, and concentration in the serum, *Br. J. Cancer,* 37, 268, 1978.
8. **de Silva, J. A. F., Munno, N., and Strojny, N.,** Absorptiometric, polarographic, and gas chromatographic assays for the determination of N-1-substituted nitroimidazoles in blood and urine, *J. Pharm. Sci.,* 59, 201, 1970.
9. **Urtasun, R. C., Chapman, J. D., Band, P., Rabin, H. R., Fryer, C. G., and Sturmwind, J.,** Phase I study of high-dose metronidazole: a specific in vivo and in vitro radiosensitizer of hypoxic cells, *Ther. Radiat.,* 117, 129, 1975.
10. **Brooks, M. A., d'Arconte, L., and de Silva, J. A. F.,** Determination of nitroimidazoles in biological fluids by differential pulse polarography, *J. Pharm. Sci.,* 65, 112, 1976.
11. **Deutsch, G., Foster, J. L., McFadzean, J. A., and Parnell, M.,** Human studies with "high dose" metronidazole: a non-toxic radiosensitizer of hypoxic radiosensitizer of hypoxic cells, *Br. J. Cancer,* 31, 75, 1975.
12. **Kane, P. O.,** Polarographic methods for the determination of two antiprotozoal nitroimidazole derivatives in materials for biological and non-biological origin, *J. Polarogr. Soc.,* 7, 58, 1961.
13. **Welling, P. G. and Monroe, A. M.,** The pharmacokinetics of metronidazole and tinidazole in man, *Arzneim.-Forsch.,* 22, 2128, 1972.
14. **Marques, R. A., Stafford, B., Flynn, N., and Sadee, W.,** Determination of metronidazole and misonidazole and their metabolites in plasma and urine by high-performance liquid chromatography, *J. Chromatogr.,* 146, 163, 1978.
15. **Lanbeck, K. and Lindstrom, B.,** Determination of metronidazole and tinidazole in plasma and feces by high-performance liquid chromatography, *J. Chromatogr.,* 162, 117, 1979.
16. **Little, C. J., Dale, A. D., and Evans, M. B.,** "C22" — a superior bonded silica for use in reverse-phase high-performance liquid chromatography, *J. Chromatogr.,* 153, 543, 1978.
17. **Workman, P., Little, C. J., Marten, T. R., Dale, A. D., Ruane, R. J., Flockhart, I. R., and Bleehen, N. M.,** Estimation of the hypoxic cell-sensitizer misonidazole and its O-demethylated metabolite in biological materials by reversed-phase high-performance liquid chromatography, *J. Chromatogr.,* 145, 507, 1978.

Chapter 29

# MORPHINE IN PLASMA BY LIQUID CHROMATOGRAPHY

**Kristi Klippel**

## TABLE OF CONTENTS

I. Introduction ................................................................. 154

II. Principle ..................................................................... 154

III. Materials and Methods ..................................................... 154
    A. Equipment ............................................................. 154
    B. Reagents ............................................................... 154
    C. Standards .............................................................. 155
    D. Procedure ............................................................. 155
    E. Calculations .......................................................... 155

IV. Results ...................................................................... 155
    A. Optimization of Chromatography .................................. 155
    B. Linearity ............................................................... 155
    C. Recovery .............................................................. 155
    D. Interferences ......................................................... 155

V. Comments .................................................................. 156

References ........................................................................ 156

## I. INTRODUCTION

Morphine is well-known as a potent narcotic analgesic because of its effectiveness against all types of acute or severe pain. The mechanism of its action is not exactly known, but the effects are believed to be due to the depressant action of the drug on areas of the central nervous system. The sedative effects and decrease of the anxiety level serves to increase the tolerance to pain, decrease the patient's perception of his surroundings, and create a feeling of euphoria. Other uses of morphine include administration as a preanesthetic sedative and as a respiratory depressant and anticonvulsant in cases of pulmonary edema, cough, and labored breathing.

Various methods of detecting morphine can be found in the literature. GCMS and GLC using flame ionization or electron capture detection all involve derivatization of morphine with varied results. Flame ionization detection has limited applicability due to low sensitivity whereas electron capture detection can have higher sensitivity but only if certain derivatives are used.[1-3] Radioimmunoassay and enzyme immunoassay have solved the problem of sensitivity but interferences due to metabolites or structurally similar compounds limit selectivity.[4-7]

In recent years, several methods using high pressure liquid chromatography (HPLC) have been published.[8-9] Because morphine is well-suited for the reverse-phase column, separation from other opiates can easily be achieved. Various detection methods have been used, such as UV and fluorescence, but these are limited by either low sensitivity or tedious sample preparation. Electrochemical detection provides the sensitivity necessary for analysis of plasma levels of morphine, limits interferences due to the low oxidation potential required, and incorporates a fast, effective sample work-up.

## II. PRINCIPLE

After adjusting the pH of the serum with a phosphate buffer, morphine is extracted from the matrix with chloroform/isopropanol solvent. The sample is centrifuged to separate the phases, an aliquot of the organic layer is removed, evaporated to dryness under nitrogen, reconstituted in mobile phase, and injected. The separation is achieved on a reverse phase $C_{18}$ column using sodium perchlorate/sodium citrate buffer-acetonitrile as the mobile phase. Morphine is detected electrochemically at a potential setting of +700 mV on a glassy carbon electrode.

## III. MATERIALS AND METHODS

### A. Equipment
An LC-154T liquid chromatograph (Bioanalytical Systems Inc., W. Lafayette, Ind.) or equivalent should be used in this procedure. This system consists of an LC-4B/17 electrochemical detector with a TL-5A glassy carbon working electrode, a LC-22A/23A temperature controller, a Biophase® ODS 5 μm reverse-phase $C_{18}$ column, a 7125 injection valve, and an RYT strip-chart recorder.

Other necessary equipment includes a vortex mixer (Scientific Products, McGaw Park, Ill.), centrifuge (Bioanalytical Systems), nitrogen evaporator (Organomation Associates, Northborough, Mass.), and 15 mℓ nalgene centrifuge tubes (Scientific Products).

### B. Reagents
**Acetonitrile** — Baker "Resi-Analyzed" (W. T. Baker Chemical Co., Philadelphia, Pa.)
**Chloroform** — SpectrAR® (Mallinkrodt, Paris, Ky.)

**Isopropanol** — SpectrAR® (Mallinkrodt)

**Mobile phase** — 100 mℓ Acetonitrile and 900 mℓ 0.2 $M$ sodium perchlorate per 0.005 $M$ sodium citrate buffer, pH 5.0. Filter through an 0.2 μm membrane filter (Rainin Instruments, Woburn, Mass.) before use.

**Sodium perchlorate/sodium citrate buffer** — Add 28.1 g sodium perchlorate and 1.47 g sodium citrate to 1 ℓ distilled water. Adjust pH to 5.0 using concentrated HCl.

**0.1 $M$ Phosphate buffer** — Dissolve 0.71 g $Na_2HPO_4$ in 50 mℓ distilled water. Adjust pH to 8.9.

**Chloroform/isopropanol solvent** — Mix 900 mℓ chloroform and 100 mℓ isopropanol.

### C. Standards

A concentrated morphine stock standard is prepared by dissolving 15 mg morphine in 20 mℓ methanol and diluting to 50 mℓ with distilled water. This should be diluted appropriately to desired concentration for working standards. All standards should be stored at 0°C.

### D. Procedure

Add 125 μℓ 0.1 $M$ phosphate buffer and 200 μℓ plasma to a 15 mℓ centrifuge tube and vortex. Add 5 mℓ chloroform/isopropanol, cap tubes, and vortex 120 sec. Centrifuge 5 min and remove white upper aqueous layer with disposable pipette. Transfer 4 mℓ of the organic layer to a clean nalgene tube and evaporate to dryness under nitrogen at 40°C. Reconstitute sample in 200 μℓ of mobile phase and inject 50 μℓ.

### E. Calculations

Quantification was achieved using peak heights. The concentration was calculated by interpolation on a calibration curve for morphine standards.

## IV. RESULTS

### A. Optimization of Chromatography

Morphine was eluted from the reverse phase column with the mobile phase previously described at a flow rate of 1.9 mℓ/min, yielding an elution time of 5.8 min. The temperature of the column was maintained at 30°C using the temperature controller. The compound was detected at a potential of +0.70 V vs. Ag/AgCl on a glassy carbon working electrode. A cell configuration employing a glassy carbon auxiliary electrode located immediately opposite the working electrode was used to compensate for any potential shift due to iR drop.

### B. Linearity

The assay is linear from 0 to 400 ng/mℓ morphine injected in both aqueous standards and spiked plasma samples.

### C. Recovery

The recovery of morphine for the above procedure was 92.85 ± 5.95% (n = 8). The minimum detectable concentration was 0.5 of ng of injected morphine, a plasma concentration of 10 ng/mℓ, with a signal-to-noise ratio of 3.

### D. Interferences

In both this procedure and liquid chromatography/electrochemistry (LCEC) methodology references cited[8,9] no interfering peaks with morphine were seen in the plasma extracts.

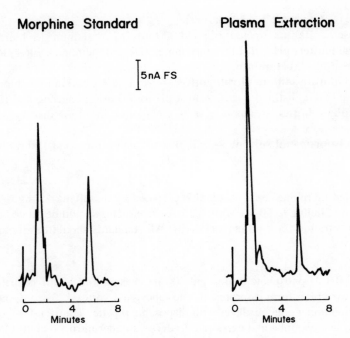

FIGURE 1. Chromatograms for morphine under conditions outlined in text. (A) Standard solution and (B) spiked plasma (50 ng/m$\ell$) carried through extraction procedure.

## V. COMMENTS

Nalgene tubes were used for this procedure because low concentrations of morphine (approximately 10 ng) are known to be readily adsorbed to active sites on unsilanized glassware.[10] Care must be taken to remove all of the white aqueous layer after centrifugation of the chloroform extraction step, including the film which forms at the interface of the two phases. If this is not sufficiently removed, the resulting chromatograms have many interfering peaks and morphine cannot be resolved.

The sensitivity of this method is more than adequate for the measurement of therapeutic levels of morphine, which are accepted as being from a few nanograms to 100 ng/m$\ell$[8,10] (Figure 1).

## REFERENCES

1. **Hipps, P. P., Eveland, M. R., Mayer, E. R., Sherman, W. R., and Cicero, T. J.**, Mass fragmentography of morphine: relationship between brain levels and analgesic activity, *J. Pharmacol. Exp. Ther.*, 196, 642, 1976.
2. **Rasmussen, K. E.**, Quantitative morphine assay by means of gas-liquid chromatography and on-column silylation, *J. Chromatogr.*, 120, 491, 1976.
3. **Nicolau, G., van Lear, G., Kaul, B., and Davidow, B.**, Determination of morphine by election capture gas-liquid chromatography, *Clin. Chem.*, 23, 1640, 1977.
4. **Spector, S. and Parker, C. W.**, Morphine: radioimmunoassay, *Science*, 168, 1347, 1970.
5. **van Vunakis, H., Wasserman, E., and Levine, L.**, Specificities of antibodies to morphine, *J. Pharmacol. Exp. Ther.*, 180, 514, 1972.

6. **Rubenstein, K. R., Schneider, R. S., and Ullman, E. F.,** Homogeneous enzyme immunoassay, a new immunochemical technique, *Biochem. Biophys. Res. Commun.,* 47, 846, 1972.
7. **Rowley, G. L., Rubenstein, K. E., Huisjin, J., and Ullman, E. F.,** Mechanism by which antibodies inhibit napten-malate dehydrogenase conjugates. An enzyme immunoassay for morphine, *J. Biol. Chem.,* 250, 3759, 1975.
8. **White, M. W.,** Determination of morphine and its morphine-3-glucoronide, in blood by high-performance liquid chromatography with electrochemical detection, *J. Chromatogr.,* 178, 229, 1979.
9. **Wallace, J. E., Harris, S. C., and Peek, M. W.,** Determination of morphine by liquid chromatography with electrochemical detection, *Anal. Chem.,* 52, 1328, 1980.
10. **Sadee, W. and Beelen, G. C. M.,** *Drug Level Monitoring,* John Wiley and Sons, New York, 1980, 343.

Chapter 30

# THEOPHYLLINE BY ULTRAVIOLET DETECTION

## George R. Gotelli and Jeffrey H. Wall

## TABLE OF CONTENTS

I. Introduction ................................................................. 160

II. Principles .................................................................. 160

III. Materials and Methods ....................................................... 160
    A. Equipment ............................................................ 160
    B. Reagents ............................................................. 160
    C. Standards ............................................................ 160
    D. Procedure ............................................................ 160
    E. Calculations ......................................................... 161

IV. Results .................................................................... 162
    A. Optimization of Chromatography ....................................... 162
    B. Linearity ............................................................ 162
    C. Recovery ............................................................. 162
    D. Reproducibility ...................................................... 162
    E. Interferences ........................................................ 162
    F. Accuracy ............................................................. 162

V. Comments ................................................................... 162

Reference ....................................................................... 162

## I. INTRODUCTION

Theophylline is a broncodilating agent used in the management of asthma. Its blood concentration is frequently measured as a guide in attaining therapeutic levels while avoiding toxicity and thus a specific assay method is required. UV spectrophotometric methods are not specific while gas-liquid chromatographic methods usually require tedious extractions and derivitizations. Liquid chromatographic methods are simple, specific, and more sensitive than other methods of analysis (Figure 1).

## II. PRINCIPLE

Theophylline and added internal standard are removed from serum by adsorption onto octadecylsilane extraction columns. After washing the column with water, the theophylline and the internal standard are eluted with methanol and an aliquot of the methanol is injected onto a reversed-phase C-18 liquid chromatographic column. The drugs are detected at 254 nm and quantitated by peak height ratio.

## III. MATERIALS AND METHODS

### A. Equipment
A liquid chromatography (LC) system equivalent to the following is recommended. A Series 1 pump (Perkin-Elmer Corp., Norwalk, Conn.), a Rheodyne® 7105 valve (Rheodyne, Cotati, Calif.), a temperature controlled oven (Perkin-Elmer LC-100), a reversed-phase C-18 column (Altex Ultrasphere® ODS, 15 cm × 4.6 mm i.d. Altex Scientific, Berkeley, Calif.) and a fixed wavelength detector (Perkin-Elmer LC-15) with a 10 mV strip-chart recorder (Perkin-Elmer 123).

### B. Reagents
The mobile phase consists of 7% methanol and 3.5% acetonotrile in 5 mmol/$\ell$ potassium dihydrogen phosphate buffer, pH 4.4 The octadecylsilane extraction columns (Bond-Elut® extraction columns) are from Analytichem International, Inc., Harbor City, Calif.

### C. Standards
The theophylline and the internal standard (β-hydroxyethyltheophylline) are from Sigma Chemical Co., St. Louis. The working internal standard consists of 25 μg/m$\ell$ β-hydroxyethyltheophylline in 0.1 mol/m$\ell$ potassium dihydrogen phosphate buffer, pH 4.4. The working drug reference standard consists of 25 μg/m$\ell$ each theophylline and β-hydroxyethyltheophylline in methanol.

### D. Procedure
*Note:* The Bond-Elut® extraction columns work most efficiently when using the Vac-Elut® vacuum chamber supplied by Analytichem International.

Attach vacuum to vacuum chamber and wash each extraction column with 2 column volumes of methanol followed by 2 column volumes of water. Disconnect vacuum and add 100 μ$\ell$ of working internal standard and 100 μ$\ell$ of unknown to the columns. Connect vacuum and pass two column volumes of water through the column. Disconnect the vacuum and place an eluate collection tube inside the vacuum chamber. Add 300 μ$\ell$ of methanol to columns and apply vacuum. Collect the methanol eluate, disconnect the vacuum, and mix the eluate.

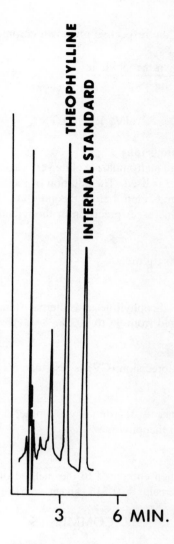

FIGURE 1. Chromatogram of a sample containing 16 μg of theophylline per milliliter of blood (obtained in the authors' laboratory).

*Note:* The extraction columns may be washed for reuse by passing two columns of methanol and 2 vol of water through the column. They can be reused up to seven times.

Inject 20 μℓ of the methanol eluate onto the column at a mobile phase flow rate of 3.0 mℓ/min. The column is maintained at 50°C and the detector is set at 254 nm. Inject 20 μℓ of the drug reference standard onto the liquid chromatographic column using the same conditions.

### E. Calculations

Calculate the response factor (RF) from the drug reference standard chromatogram.

$$\frac{\text{peak height of Int. Std.}}{\text{peak height of theophylline}} = \text{Response Factor (RF)}$$

Calculate the unknowns from the respective unknown chromatograms.

$$\frac{\text{peak height of unknown theophylline}}{\text{peak height of Int. Std.}} \times RF \times 25 = \mu g/m\ell \text{ theophylline}$$

## IV. RESULTS

### A. Optimization of Chromatography

Several thiazide diuretics and metronidazole elute very close to theophylline when a mobile phase of acetonitrile and water is used. The addition of methanol changes the selectivity of the system and these compounds elute before theophylline. The 50°C column temperature decreases back pressure and increases precision of the retention times.

### B. Linearity

The method is linear to 100 $\mu g/m\ell$.

### C. Recovery

Absolute recovery of both theophylline and internal standard exceeds 95%. Analytical recovery of theophylline ranged from 93 to 102%.

### D. Reproducibility

Day-to-day and within-run precision (CV) is less than 5%.

### E. Interferences

The only known interference is 1,7 dimethylxanthine, a minor metabolite of caffeine. This compound coelutes with theophylline.

### F. Accuracy

The regression analysis when compared to the method of Kabra and Marton[1] was r = 0.995, slope = 0.978, y-intercept = 0.516, and n = 51.

## V. COMMENTS

Extraction of the samples on the extraction columns is simple, rapid, and results in very clean eluates for injection. Analytical columns have lasted over 16 months or 4000 samples. Very few unidentified interferences have been encountered.

## REFERENCE

1. **Kabra, P. K. and Marton, L. J.**, Liquid chromatographic analysis for serum theophylline in less than 70 seconds, *Clin. Chem.*, 28, 687, 1982.

Chapter 31

# VERY HIGH SPEED LIQUID CHROMATOGRAPHIC ANALYSIS OF THEOPHYLLINE IN SERUM

**Pokar M. Kabra**

## TABLE OF CONTENTS

I. Principle ................................................................. 164

II. Materials and Methods ................................................. 164
    A. Equipment .......................................................... 164
    B. Reagents ............................................................ 164
    C. Standards ........................................................... 164
    D. Procedure ........................................................... 164
    E. Calculations ........................................................ 165

III. Results ................................................................. 165
    A. Linearity ........................................................... 165
    B. Recovery ........................................................... 165
    D. Interference ........................................................ 165
    E. Accuracy ........................................................... 165

IV. Comments .............................................................. 166

References ................................................................. 166

## I. PRINCIPLE

Theophylline is extracted from 100 μℓ of serum containing β-hydroxyethyl theophylline as an internal standard with a Bond-Elut® column. The Bond-Elut® column is eluted with 300μℓ of methanol and an aliquot of the eluate is injected onto a reversed-phase high speed column. Theophylline and internal standard are eluted with acetonitrile/phosphate buffer mobile phase. The drugs are detected by their absorption at 273 nm and quantitated by peak height or peak area ratios.

## II. MATERIALS AND METHODS

### A. Equipment
A liquid chromatograph (LC) system equivalent to the following is recommended. A Model Series 2 or Series 3 liquid chromatograph equipped with a Model LC-100 column oven, a Model LC-85 variable wavelength detector, and a Sigma® 10 Data System (all from Perkin Elmer Corp., Norwalk, Conn.), can be used. A high speed chromatography package consisting of a Rheodyne® Model 7125 injector with a 6μℓ loop, a Perkin-Elmer 2.4 μℓ micro flow cell for the LC-85 detector, and 80 cm × 0.18 mm (i.d.) stainless-steel connecting tubing is used to minimize extra-column band broadening. A Perkin-Elmer column (125 × 4.6 mm, i.d.) column packed with 5 μm particle size $C_{18}$ reversed-phase packing is mounted in the oven.

### B. Reagent
All chemicals used are of reagent grade. Acetonitrile, methanol, all distilled in glass, can be obtained from Burdick and Jackson Laboratories, Inc., Muskegon, Mich.

Phosphate buffer, 20 mmol/ℓ, pH 3.6, is prepared by dissolving 2.7 g of $KH_2PO_4$ in 1 ℓ of water. The pH is adjusted to 3.6 with phosphoric acid. Mobile phase: 95 mℓ of acetonitrile in 905 mℓ of 20 mmol/ℓ phosphate buffer.

### C. Standards
Theophylline and β-hydroxyethyltheophylline (internal standard) are from Sigma Chemical Co., St. Louis. A standard mixture for chromatography is prepared by dissolving 25 mg each of theophylline and internal standard in 100 mℓ of methanol. The solution is stable at 4°C for at least 1 year. The stock internal standard is prepared by dissolving 25 mg of β-hydroxyethyltheophylline in 100 mℓ of water. The working internal standard is prepared by diluting the stock internal standard tenfold with 0.1 mol/ℓ $KH_2PO_4$ buffer (pH 4.4).

### D. Procedure
Place Bond-Elut® columns on top of the Vac-Elut® chamber. Connect the Vac-Elut® chamber to vacuum and pass two column volumes of methanol and distilled water through each column. After disconnecting the vacuum, place 100 μℓ of working internal standard (25 μg/mℓ of β-hydroxyethyltheophylline) and 100 μℓ of standard, control, or patient serum into each labeled column. Connect the Vac-Elut® chamber to vacuum and wash each column with 2 vol of distilled water. Disconnect vacuum and place a rack containing labeled 10 × 75 mm glass tubes in the Vac-Elut® chamber. Pipette 300 μℓ of methanol onto each column and connect to vacuum. Remove the rack of tubes from the vacuum chamber, shake the tubes to mix the eluate, and inject 6μℓ into the liquid chromatograph. The column, which is maintained at 50°C, is eluted with acetonitrile/phosphate buffer (9.5/90.5 by vol) at the rate of 4.5 mℓ/min, and the column effluent is monitored at 273 nm (Figure 1).

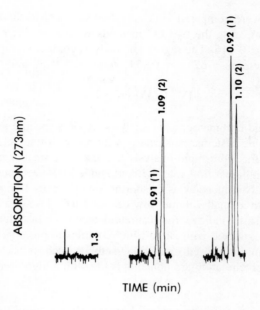

FIGURE 1. Chromatogram of (left) theophylline-free serum; (middle) a patient's serum containing 5.6 mg of theophylline per liter; (right) a patient's serum containing 21.5 mg of theophylline per liter. (1) theophylline, (2) β-hydroxyethyltheophylline (internal standard). (From Kabra, P. M. and Marton, L. J., *Clin. Chem.*, 28, 687, 1982. With permission.)

### E. Calculations

1. Serum standards are carried through the procedure and used to prepare a standard curve based on relative peak height or peak area ratios.
2. Calculate the concentration of theophylline in the specimen by direct comparison with the data obtained from the standard curve of serum standards.

## III. RESULTS

### A. Linearity
The method is linear up to 50.0 mg/ℓ, with minimum sensitivity of 0.5 mg/ℓ.

### B. Recovery
Absolute recovery of both theophylline and internal standard exceeds 95%. Analytical recovery of theophylline ranges from 97 to 102%.

### C. Precision
Both within-day and day-to-day precision are less than 4% coefficient of variation.

### D. Interference
Of more than 60 drugs tested for interference, only dyphylline eluted close to theophylline and partially interfered with theophylline analyses.

### E. Accuracy
To assess the accuracy of the method, results from more than 100 samples from subjects

receiving theophylline were compared with an established LC method[2] or immunoassay.[3] The regression data comparing the two LC methods were n = 51, r = 0.995, slope = 0.978, and y-intercept = 0.516. The regression analysis comparing high-speed LC method and the EMIT assays were: n = 67, r = 0.971, slope = 0.925, and y-intercept = 0.758.

## IV. COMMENTS

Very high-speed liquid chromatographic analysis requires specific modifications of conventional chromatographic systems, particularly with regard to the column and detector. A shorter column allows for more rapid analysis because of a smaller column void volume and the ability to elute at higher flow rates without undue back-pressure. Extra-column band broadening is significantly reduced by using shorter, small bore (0.18 mm) connecting tubing and an UV detector with a small volume flow cell (2.4 $\mu\ell$). Use of a detector with a very fast response time (135 msec) allows for accurate detection of fast eluting analytes. Serum or plasma samples are extracted with Bond-Elut® columns to prolong the useful life of the analytical column. Commonly used extraction methods such as acetonitrile or methanol techniques for protein precipitation are noncompatible with a high-speed liquid chromatography system.

## REFERENCES

1. **Kabra, P. M. and Marton, L. J.**, Liquid-chromatographic analysis for serum theophylline in less than 70 seconds, *Clin. Chem.*, 28, 687, 1982.
2. **Gotelli, G. R., and Wall, J.**, Theophylline by ultraviolet detection, in *Clinical Liquid Chromatography*, Kabra, P. M. and Marton, L. J., Eds., CRC Press, Boca Raton, Fla., 1984, chap. 30.
3. **Kampa, I. S., Dunikoski, L. K., Jr., Jarzabek, J. I., et al.**, Comparison of three assay procedures for theophylline determination, *Ther. Drug. Monit.*, 1, 249, 1979.

Chapter 32

# SIMULTANEOUS REVERSED-PHASE LIQUID CHROMATOGRAPHIC ANALYSIS OF AMITRIPTYLINE, NORTRIPTYLINE, IMIPRAMINE, DESIPRAMINE, DOXEPIN, AND NORDOXEPIN

**Pokar M. Kabra**

## TABLE OF CONTENTS

I. Introduction ................................................................. 168

II. Principle ..................................................................... 168

III. Materials and Methods ..................................................... 168
    A. Equipment ............................................................ 168
    B. Reagents .............................................................. 168
    C. Standards ............................................................. 169
    D. Procedure ............................................................. 170
    E. Calculations .......................................................... 170

IV. Results ....................................................................... 170
    A. Linearity .............................................................. 170
    B. Recovery .............................................................. 170
    C. Precision ............................................................. 170
    D. Sensitivity and Detection Limit ..................................... 170
    E. Interferences ......................................................... 170

V. Comments .................................................................... 170

VI. Interpretation ............................................................... 171

References ........................................................................ 171

## I. INTRODUCTION

The use of tricyclic antidepressant drugs for the treatment of depression is increasing.[1] The tricyclic antidepressant drugs commonly prescribed in the U.S. include the tertiary amines, amitriptyline, imipramine, and doxepin, and the secondary amines, nortriptyline, desipramine, and protriptyline. The tertiary amines are metabolized in part by N-demethylation to pharmacologically active secondary amines. Therefore, any method used for measurement of tricyclic antidepressant concentration after the administration of tertiary amines, must be capable of measuring both the tertiary and secondary amines. The correlation between plasma levels of some tricyclic antidepressants and therapeutic effects suggests that plasma levels may provide valuable information in improving the clinical management of depressed patients.[2] The determination of plasma levels may also be useful in monitoring compliance and toxic effects.

The tricyclic drugs have been determined using a variety of analytical techniques including UV spectroscopy, fluorescence spectrophotometry, and thin layer chromatography.[3] Procedures based on these techniques are time consuming, lack adequate sensitivity, and are prone to interference. Gas liquid chromatography (GLC) or liquid chromatography (LC) are most frequently used to assay tricyclic antidepressant drugs. Many GLC methods use flame ionization detectors, which can only detect high therapeutic or toxic drug concentrations.[3] Gas chromatography-mass spectrometry[4] and gas chromatography with alkaline-flame detection[5] have provided the most sensitive methods for monitoring plasma levels of tricyclic antidepressant drugs and their metabolites, although many of these procedures involve a time consuming sample preparation and derivatization step prior to the chromatography. Normal phase[6] and reversed-phase liquid chromatographic[7] methods with and without ion-pairing have been reported for the analysis of tricyclic antidepressant drugs. Because of the basic nature of tricyclic antidepressant drugs, they do not chromatograph well on normal phase or reversed-phase columns, therefore, high pH mobile phases or ion-pair techniques have been used. A reversed-phase liquid chromatographic method in which n-nonylamine is added to the mobile phase to reduce adsorption of the tricyclic antidepressant drugs on to the reversed-phase column is presented here.[8]

## II. PRINCIPLE

The tricyclic antidepressant drugs and the internal standard (loxapine) are extracted from 2 m$\ell$ of serum at alkaline pH into butylchloride and back extracted into 200 µ$\ell$ of 0.025 mol/$\ell$ HCl. An aliquot of the acid phase is injected onto a reversed-phase column and the drugs are eluted with acetonitrile/phosphate buffer, pH 3.2 (21/79, by vol), containing 0.6 m$\ell$ of nonylamine per liter of phosphate buffer. The drugs are detected by their absorption at 200 nm and quantitated from their peak heights (Figure 1).

## III. MATERIALS AND METHODS

### A. Equipment

A liquid chromatograph system equivalent to the following is recommended. A Model Series 2 or Series 3 liquid chromatograph equipped with a Model LC75 variable wavelength detector, a Rheodyne® 7105 valve (all from Perkin-Elmer Corp., Norwalk, Conn.), and a reversed-phase octadecyl column 15 cm × 4.6 mm (Ultrasphere®-ODS, Altex Scientific, Berkeley, Calif.), a strip-chart recorder or a digital data system (Sigma® 10, Perkin Elmer).

### B. Reagents

All chemicals are of reagent grade. Acetonitrile and butylchloride, distilled in glass, UV

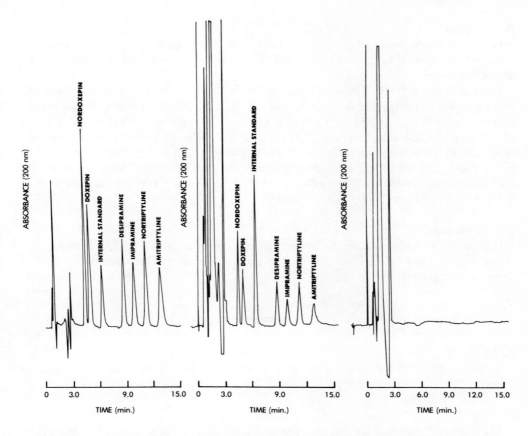

FIGURE 1. Left: Chromatogram of a standard mixture of tricyclic drugs, middle: chromatogram of a serum supplemented with approximately 80 ng of each tricyclic drug, except for the internal standard, which is 200 ng; right: chromatogram of a drug-free serum. (From Kabra, P. M., Mar, N. A., and Marton, L. J., *Clin. Chim. Acta,* 111, 123, 1981. With permission.)

grade, can be obtained from Burdick and Jackson, Muskegon, Mich. *n*-Nonylamine is from Aldrich Chemical Co., Milwaukee, Wis.

Phosphate buffer: approximately 10 mmol/ℓ, containing *n*-nonylamine is prepared by dissolving 1.34 g of $KH_2PO_4$ to 1L of distilled water, followed by 0.6 mℓ of *n*-nonylamine. This solution is titrated to pH 3.0 with phosphoric acid. Mobile phase: acetonitrile/phosphate buffer, 21/79 by volume, final pH 3.2

## C. Standards

All drug standards (amitriptyline, nortriptyline, imipramine, desipramine, doxepine, nordoxepin, and loxapine) were gifts from the Institute of Forensic Sciences, Oakland, Calif. A stock standard mixture is prepared as follows: 25 mg each of doxepin, nordoxepin, imipramine, desipramine, amitriptyline, and nortriptyline are dissolved in 100 mℓ of methanol. This solution is stable at 4°C for at least 6 months. Working serum standards and controls are made by adding various amounts of these drugs to drug free serum. Working aliquots of these standards and controls are frozen at −20°C and found to be stable for at least 6 months. Stock internal standard is prepared by dissolving 25 mg of loxapine in 100 mℓ of methanol. Working internal standard is prepared by diluting 4 mℓ of stock solution to 100 mℓ with 0.1 mol/ℓ HCl.

### D. Procedure

Glassware preparation: all glassware should be acid washed before use. All centrifuge tubes are rinsed with hexane and methanol. These steps are necessary to eliminate glass adsorption losses.

*Method:* transfer 2 mℓ of serum or plasma into a 12 mℓ glass centrifuge tube, add 100 μℓ of loxapine as internal standard, 200 μℓ of 1.5 mol/ℓ NaOH, and 10 mℓ of butylchloride. Rotate mix for 5 min, centrifuge for 5 min at 500 × g, and transfer organic phase into a 12 mℓ glass conical tube containing 200 μℓ of 0.025 mol/ℓ HCl. Shake the mixture mechanically for 10 min and centrifuge for 2 min at 500 × g. After discarding the top organic layer, inject 100 μℓ of the aqueous phase into the chromatograph using the following chromatographic conditions: flow rate — 2.5 mℓ/min; detector — UV, 200 nm; temperature ambient; and sensitivity — 0.03 A full scale.

### E. Calculations

Each drug can be quantitated by measuring the ratio of peak height of that drug to that of internal standard in the unknown sample and comparing it with a serum standard of known concentration.

## IV. RESULTS

### A. Linearity

A linear relationship between peak height ratios and drug concentrations is obtained between 22 to 1780μg/ℓ.[8]

### B. Recovery

The analytical recovery for the six tricyclic antidepressants range from 73 to 107%.[8]

### C. Precision

The precision for tricyclic antidepressant drugs equals less than 8% coefficient for variation for both within-day and day-to-day analysis.[8]

### D. Sensitivity and Detection Limit

Under the conditions of assay, a detection limit of 5 μg/ℓ can be achieved with a signal to noise ratio exceeding three. Sensitivity of 2 to 3 μg/ℓ may be achieved by injecting a larger sample volume.[8]

### E. Interferences

A total of 35 commonly used basic and neutral drugs that may be partially extracted into the acidic aqueous extract were tested for possible interference. Propoxyphene is not completely resolved from nortriptyline and may result in falsely elevated results for nortriptyline. Commonly used benzodiazepines and phenothiazines do not interfere with the assay.[8]

## V. COMMENTS

The extraction procedure was optimized for optimum recovery of these drugs by elimination of the usual evaporation and reconstitution steps, reduction of the sample manipulations steps, and pretreatment of all glassware to reduce adsorption losses. The two most important factors influencing the separation of tricyclic drugs are the pH of the mobile phase and the addition of *n*-nonylamine as a competing base for free silica sites. This is crucial in separating these drugs and in improving the peak symmetry.

## VI. INTERPRETATION

The commonly accepted therapeutic ranges for these drugs are listed below.[9]

| Drug | Therapeutic concentration |
|------|---------------------------|
| Amitriptyline | 125-250 µg/ℓ |
| Nortriptyline | 50-150 µg/ℓ |
| Imipramine | 150-300 µg/ℓ |
| Desipramine | Not well-established |
| Doxepin | Not well-established |

## REFERENCES

1. **Dito, W. R.**, Tricyclic antidepressants, *Diagn. Med.*, 5, 48, 1979.
2. **Scoggins, B. A., Maguire, K. P., Norman, T. R., and Burrows, G. D.**, Measurement of tricyclic antidepressants. II. Applications of methodology, *Clin. Chem.*, 26, 805, 1980.
3. **Schmidt, G. J.**, Tricyclic antidepressants, in *Liquid Chromatography in Clinical Analysis*, Kabra, P. M. and Marton, L. J., Eds., Humana Press, Clifton, N.J., 1981, 187.
4. **Chinn, D. M., Jennison, T. A., Crouch, D. J., Peak, M. A., and Thatcher, G. W.**, Quantitative analysis for tricyclic antidepressant drugs in plasma or serum by gas chromatography chemical ionization mass spectrometry, *Clin. Chem.*, 26, 1201, 1980.
5. **Bailey, D. N. and Jatlow, P. I.**, Gas-chromatographic analysis of therapeutic concentrations of amitriptyline and nortriptyline in plasma, with use of a nitrogen detector, *Clin. Chem.*, 22, 777, 1976.
6. **Vandemark, F. L., Adams, R. F., and Schmidt, G. J.**, Liquid-chromatographic procedure for tricyclic drugs and their metabolites in plasma, *Clin. Chem.*, 24, 87, 1978.
7. **Proelss, H. F., Lohmann, H. J., and Miles, D. G.**, High-performance liquid chromatographic simultaneous determination of commonly used tricyclic antidepressants, *Clin. Chem.*, 24, 1948, 1978.
8. **Kabra, P. M., Mar, N. A., and Marton, L. J.**, Simultaneous liquid chromatographic analysis of amitriptyline, nortriptyline, imipramine, desipramine, doxepin, and nordoxepin, *Clin. Chim. Acta*, 111, 123, 1981.
9. **Jatlow, P.**, Measurement of tricyclic antidepressant drugs, in *Methodology for Analytical Toxicology*, Vol. 2, Sunshine, I. and Jatlow, P., Eds., CRC Press, Boca Raton, Fla., 1982, 119.

Chapter 33

# DETERMINATION OF AMITRIPTYLINE, IMIPRAMINE, NORTRIPTYLINE, AND DESIPRAMINE

**Gary J. Schmidt**

## TABLE OF CONTENTS

I. Introduction .................................................................... 174

II. Principle ....................................................................... 174

III. Methods and Materials ......................................................... 174
    A. Equipment ............................................................... 174
    B. Reagents ................................................................. 174
    C. Standards ................................................................ 174
    D. Procedure ................................................................ 175
    E. Calculation .............................................................. 175

IV. Results ......................................................................... 175
    A. Linearity ................................................................. 175
    B. Recovery ................................................................. 175
    C. Detection ................................................................ 175
    D. Precision ................................................................ 175
    E. Interferences ............................................................ 175
    F. Patient Serum ........................................................... 176

V. Comment ....................................................................... 178

Reference .......................................................................... 178

## I. INTRODUCTION

The tricyclic drugs are an important and widely used class of antidepressants. The clinical usefulness of determining serum concentrations of these drugs is still under debate. However, indications are that their determination is useful both to control overdosage and to monitor clinical efficacy. A variety of analytical techniques are used for determining these drugs including gas chromatography, liquid chromatography (LC), and radioimmunoassay. Methods based upon LC are relatively straightforward and provide a means of determining both parent drugs and demethylated metabolites over the entire therapeutic concentration range.

## II. PRINCIPLE

The drugs are extracted from alkalinized plasma into hexane/isoamyl alcohol, 98/2 (v/v), with protriptyline being used as an internal standard. After evaporation of the extraction solvent, the residue is redissolved in 20 $\mu\ell$ of mobile phase and a 10 $\mu\ell$ aliquot is chromatographed on a 5 $\mu$m silica column using isocratic elution. The drugs are detected using an UV absorption detector set at 211 nm. Either peak height ratios or peak area ratios may be used for quantitation.

## III. MATERIALS AND METHODS

### A. Equipment

A liquid chromatography system equivalent to the following is used. A model Series 2/2 liquid chromatograph (Perkin-Elmer Corp., Norwalk, Conn.) equipped with a Rheodyne® 7125 injection valve (Rheodyne, Cotati, Calif.), a UV detector (Perkin-Elmer Model LC-75 or equivalent), a column oven (Perkin-Elmer Model LC-100), and a strip-chart recorder (Perkin-Elmer Model 56). The column is a 5-$\mu$m silica normal phase 25 cm × 4.6 mm i.d. (Perkin-Elmer Silica B/5). Special glassware includes 16 × 100 mm PTFE-lined screw-capped test tubes, 5-m$\ell$ conical centrifuge tubes, and 20-$\mu\ell$ disposable glass pipettes. A bench-top centrifuge, a heating block, an evaporation manifold with a source of dry nitrogen, and a test tube rotator rack are used.

### B. Reagents

Hexane, acetonitrile, and methanol, distilled in glass, UV grade (Burdick and Jackson Laboratories, Inc., Muskegon, Mich.) Isoamyl alcohol, redistilled before use (J. T. Baker Chemical Co., Phillipsburg, N. J.). Extraction solvent: add hexane and isoamyl alcohol in the proportions 98/2 by volume. Protriptyline, used as an internal standard, is added to provide a concentration of 100 $\mu$g/$\ell$. Sodium carbonate: prepare a saturated aqueous solution. Mobile phase: prepare by mixing 993 m$\ell$ acetonitrile and 7 m$\ell$ concentrated ammonium hydroxide. Mobile phase solutions are degassed using sonification at 90 W for 2 min.

### C. Standards

Amitriptyline, imipramine, nortriptyline, desipramine, and protriptyline as the hydrochloride salts, are obtained from U.S. Vitamin Pharmacopeial Convention, Inc., Rockville, Md. Individual stock solutions of each drug as the free base is prepared in methanol by dissolving 11.3 mg of the hydrochloride salt in 10 m$\ell$ of methanol to provide a concentration of 1 mg/m$\ell$. The solution is stored at $-20°C$.

**Chromatography test solution** — Prepared by combining 20 $\mu\ell$ of each of the five 1 mg/m$\ell$ drug stock solutions, evaporate, and reconstitute the residue with 1.0 m$\ell$ of the mobile phase to provide a concentration of 20 mg/m$\ell$. This solution is used to evaluate the chromatographic performance on a daily basis and is stored at 4°C.

**Working stock solution** — Prepared by combining 1 mℓ of the stock solution of each drug, except protriptyline, and dilute to 100 mℓ with methanol. The solution is stored at 4°C.

**Plasma Standards** — Six plasma standards are prepared from pooled drug-free plasma to provide concentrations of each drug equivalent at 0, 25, 50, 100, 250, and 500 μg/ℓ. These plasma standards are prepared by adding an appropriate volume of the working stock solution to a glass container, evaporating to dryness, and reconstituting with pooled drug-free plasma to provide the above concentrations. The plasma standards are stored frozen at −10 to −20°C.

### D. Procedure

To 2 mℓ of plasma sample or plasma standard in a 16 × 100 mm tube, add 0.5 mℓ of saturated sodium carbonate solution. Mix gently and add 5 mℓ of the extraction solvent. Rotate the tubes for 5 min (12 r/min) and then centrifuge at 2000 r/min (610 g) for 2 min. Transfer the organic (upper) layer to a 5 mℓ conical centrifuge tube and evaporate to dryness at 60°C under a stream of dry nitrogen. The residue is redissolved in 20 μℓ mobile phase and 10 μℓ is injected into the chromatograph. The mobile phase flow rate is 1.5 mℓ/min and the column temperature is 65°C. The UV detector is set at 65°C. The UV detector is set at 211 nm at a sensitivity of 0.08 absorbance units full scale. A chromatogram illustrating the separation of the four tricyclic drugs is shown in Figure 1. The retention times and relative retention as compared to protriptyline is shown in Table 1.

### E. Calculation

For each plasma standard, the peak area of the drug of interest is divided by the peak area of the protriptyline internal standard. A working curve is prepared by plotting the peak area ratio of the drug against the concentration. The patients' plasma are then analyzed, the peak area ratios calculated, and the drug concentration determined from the working curve. A separate standard curve must be prepared for each drug. Peak height ratios may also be used.

## IV. RESULTS

### A. Linearity

A linear relationship of the peak area ratios exists for each drug over the concentration range of 25 to 800 μg/ℓ.

### B. Recovery

Mean analytical recoveries for each drug are between 60 and 69% over the concentration range of 100 to 500 μg/ℓ. Incomplete recovery is compensated by the use of plasma standards.

### C. Detection

Therapeutic concentration of the drugs are determined to a sensitivity of 10 μg/ℓ.

### D. Precision

Within-run precision at a concentration of 100 μg/ℓ varied between 4.9 and 6.4% for all drugs. Day-to-day precision at this concentration varied between 5.8 and 7.9%.

### E. Interferences

The retention of drugs tested as interferences are listed in Table 2. There is a risk of interference for compounds that differ in RRT by less than 6% for relative retentions greater than 0.2 In addition, ten different drug-free plasma samples analyzed according to these

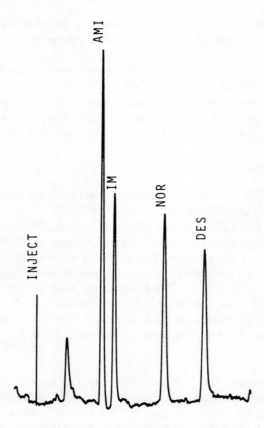

FIGURE 1. Chromatogram of a standard mixture of 50 ng of each drug. (From Vandemark, F. K., Adams, R. F., and Schmidt, G. J., *Clin. Chem.*, 24, 87, 1978. With permission.)

## Table 1
### RETENTION TIME (RT) AND RELATIVE RETENTION (RRT) OF THE TRICYCLICS

| Drug | $R_t$ (min) | $RR_t$ |
|---|---|---|
| Amitriptyline | 3.52 | 0.22 |
| Imipramine | 4.28 | 0.28 |
| Nortriptyline | 6.97 | 0.49 |
| Desipramine | 9.79 | 0.71 |
| Protriptyline | 13.51 | 1.00 |

procedures showed background values at retention times similar to that of the drugs of interest equivalent to 0 to 2 µg/ℓ.

### F. Patient Serum

Figure 2 illustrates an example of a plasma sample from a patient receiving amitriptyline. The amitriptyline concentration is 119 µg/ℓ and the nortriptyline concentration is 156 µg/ℓ.

## Table 2
## RELATIVE RETENTION TIMES OF DRUGS TESTED FOR INTERFERENCE

| Drug | RRT | Drug | RRT |
|---|---|---|---|
| Acetaminophen | 0.02 | Chlorpromazine | 0.17 |
| Phenacetin | 0.03 | Methadone | 0.17 |
| Scopolamine | 0.03 | Diphenhydramine | 0.22 |
| Cocaine | 0.07 | Doxepin | 0.22 |
| Diazepam | 0.07 | Meperidine | 0.34 |
| Phenytoin | 0.07 | Trifluoperazine | 0.39 |
| Propoxyphene | 0.11 | Perphenazine | 0.41 |
| Oxazepam | 0.12 | Prochlorperazine | 0.46 |
| Flurazepam | 0.14 | Codeine | 0.61 |
| Caffeine | 0.14 | Chlorpheniramine | 0.76 |
| Chlordiazepoxide | 0.16 | | |

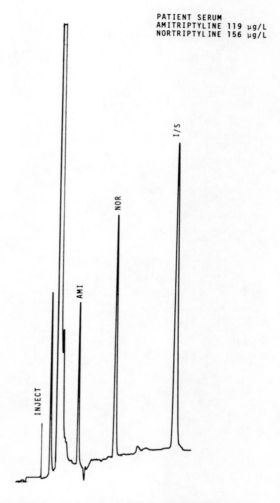

FIGURE 2. Chromatogram of a patient plasma sample containing 119 µg/ℓ amitriptyline and 156 µg/ℓ nortriptyline. (From Vandemark, F. L., Adams, R. F., and Schmidt, G. J., *Clin. Chem.*, 24, 87, 1978. With permission.)

## V. COMMENT

The therapeutic ranges are not clearly defined but appear to be in the range of 50 to 300 µg/ℓ depending upon the particular drug.

## REFERENCE

1. **Vandemark, F. L., Adams, R. F., and Schmidt, G. J.,** Liquid-chromatographic procedure for tricyclic drugs and their metabolites in plasma, *Clin. Chem.*, 24, 87, 1978.

Chapter 34

# IMIPRAMINE, DESIPRAMINE, AND METABOLITES BY LIQUID ELECTROCHEMISTRY

**Julie Morris and Ronald E. Shoup**

## TABLE OF CONTENTS

I. Introduction ................................................................. 180

II. Principle ..................................................................... 180

III. Materials and Methods ..................................................... 180
    A. Equipment .............................................................. 180
    B. Reagents ................................................................ 180
    C. Standards .............................................................. 181
    D. Procedure .............................................................. 181
    E. Calculations ........................................................... 181

IV. Results ...................................................................... 182
    A. Chromatography ..................................................... 182
    B. Linearity ............................................................... 182
    C. Recovery and Detection Limits ..................................... 182

V. Comments ................................................................... 183

References ....................................................................... 183

## I. INTRODUCTION

The tricyclic antidepressants used in the treatment of endogenous depression have been widely prescribed in recent years. Since there is a good correlation between the plasma levels of these tricyclic amines and their therapeutic efficacy, there is a need for routine laboratory procedures to assay plasma levels of tricyclic antidepressants (TCAs).

A complication in monitoring plasma levels of patients receiving TCAs is that some of the drugs produce active metabolites that have somewhat different pharmacological actions.[1] Therefore, it is necessary to monitor both TCAs and their metabolites in order to achieve effective therapeutic results.

There is some controversy as to the therapeutic ranges of the TCAs. However, most agree that 20 to 200 ng/mℓ (plasma) is a satisfactory therapeutic range. For 1 mℓ plasma samples, this requirement stipulates an analytical technique capable of very low nanogram detection limits.

## II. PRINCIPLE

The TCAs and their hydroxy metabolites are extracted from plasma based on the acid-base chracteristics of these substances. Base-buffered plasma is extracted with methyl-t-butylether, then back extracted into 0.1 $M$ HCl. The acid layer is extracted again with methyl-t-butylether. The organic layer is evaporated under $N_2$ and reconstituted in mobile phase for injection. The back extraction is effective at removing many neutral and acidic species. Detection is accomplished amperometrically, through the oxidation of the aromatic amine at a glassy carbon electrode.

## III. MATERIALS AND METHODS

### A. Equipment

A LC-154T liquid chromatograph (Bioanalytical Systems, Inc., W. Lafayette, Ind.) with a BAS LC-6 UV detector (254 nm) connected in series with an LC-4 electrochemical detector was used for all determinations. Since the mobile phase contains appreciable organic solvent, a TL-5 glassy carbon working electrode was mandated for the flowcell. A Biophase® Octyl 5 μm column (25 cm × 4.6 mm i.d., BAS) was used for all separations. A nitrogen evaporator (Organomation Associates, Shrewsbury, Mass.) and a Sorvall® GLC-2 centrifuge (DuPont Company, Newton, Conn.) were also utilized.

### B. Reagents

**Solvents** — Acetonitrile "Baker resi-analyzed" (J. T. Baker, Phillipsburg, N.J., available through American Scientific Products, McGaw Park, Ill.), methyl-t-butyl ether (Burdick and Jackson Laboratories, Inc., Muskegon, Mich.).

**0.2 $M$ Sodium perchlorate, 0.005 $M$ sodium citrate buffer** — Add 28.1 g of sodium perchlorate (Fisher Scientific) and 1.47 g of sodium citrate (Fisher Scientific, Fair Lawn, N.J.), to a 1 ℓ volumetric flask. Dilute to the mark with deionized, distilled water. Adjust the pH to 5.50 to 5.55 with glacial acetic acid. This will be used to make the working mobile phase buffer and should be made fresh at least every 2 days.

**Mobile phase** — Mix 575 mℓ of the above buffer and 325 mℓ acetonitrile in a 1 ℓ volumetric flask. Dilute with methanol to the mark. Filter through a 0.2 μm pore size membrane filter (Rainin Instruments, Woburn, Mass.) and degas prior to use. (See Table 1 for liquid chromatographic conditions).

## Table 1
## LIQUID CHROMATOGRAPHIC CONDITIONS FOR PLASMA TCA ASSAY

Liquid chromatograph: LC-154T (Bioanalytical Systems)
Mobile phase: 57.5% 0.2 $M$ NaClO$_4$, 0.005 $M$ trisodium citrate (pH 5.5)/32.5% CH$_3$CN/10% MeOH
Flow rate: 2 m$\ell$/min
Stationary phase: Biophase® Octyl 5 µm (25 cm × 4.6 mm i.d., Bioanalytical Systems)
Temperature: 30°C
UV Detector (optional): LC-6, 254 nm (Bioanalytical Systems)
EC Detector: LC-4/LC-17 detector using a TL-5 glassy carbon working electrode
Detector potential: +1.05 V (vs. Ag/AgCl)

**0.6 $M$ K$_2$CO$_3$ Buffer, pH 11.3** — Dissolve 8.3 g K$_2$CO$_3$ (Mallinckrodt, St. Louis) in 100 m$\ell$ deionized distilled water.

**0.1 $M$ HCl** — Dilute 0.83 m$\ell$ concentrated HCl (Mallinckrodt) to 100 m$\ell$ with deionized, distilled water.

### C. Standards

2-Hydroxyimipramine and 2-hydroxydesipramine were donated by Dr. Albert A. Manian, National Institute of Mental Health, Rockville, Md. Desipramine and imipramine were donated by Dr. Mark Walter, Waterbury Hospital, Waterbury, Conn.

**TCA stock Standard** — Dissolve 25 mg of each tricyclic amine in 50 m$\ell$ methanol.

**Working standards** — Dilute stock standards with appropriate volumes of mobile phase.

### D. Procedure

This procedure was adapted from that of Suckow and Cooper.[2] According to their suggestion, methyl-t-butyl ether was substituted for diethyl ether, yielding better recoveries and cleaner extracts.

To 1.0 m$\ell$ plasma, add 1.0 m$\ell$ of H$_2$O and 1.0 m$\ell$ 0.6 $M$ K$_2$CO$_3$ in a 15 m$\ell$ conical glass centrifuge tubes with screw caps. Add 8.0 m$\ell$ methyl-t-butyl ether. Mechanically shake for 5 min and centrifuge at 2000 rpm for 15 min.

Transfer the ether layer to a 15 m$\ell$ conical centrifuge tube with screw cap containing 1.2 m$\ell$ 0.1 $M$ HCl. Mechanically shake for 5 min, centrifuge at 2000 rpm for 10 min.

Aspirate the top layer to waste. Add 0.5 m$\ell$ 0.6 $M$ K$_2$CO$_3$ and 1.0 m$\ell$ methyl-t-butyl ether. Mechanically shake for 5 min, centrifuge at 2000 rpm for 5 min.

Transfer the ether layer to a 3 m$\ell$ conical test tube. Evaporate the ether under a nitrogen stream at 40°C to dryness. Reconstitute the residue in 100 µ$\ell$ of mobile phase. Inject 50 µ$\ell$.

All test tubes and centrifuge tubes were surface tested with an organosilane surface treating agent, PROSIL®-28 (PCR Research Chemicals, Gainesville, Fla., available through American Scientific Products).

### E. Calculations

Direct peak height comparisons are made between unknown plasma samples and blank plasma spiked with known amounts of standards, after processing all samples through the entire procedure. Calculations for any of the four peaks are as follows (assuming equal volumes injected):

$$\text{conc.}_{unknown} = \frac{\text{peak height (nA, unknown)}}{\text{peak height (nA, known)}} \times \text{conc.}_{known}$$

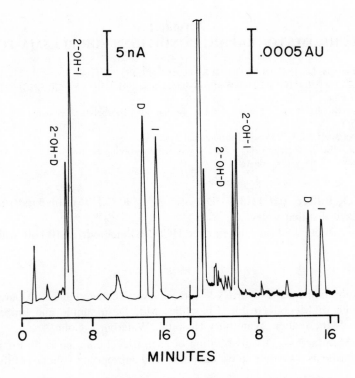

FIGURE 1. Comparison of spiked plasma chromatograms obtained by EC and UV detection. Legend: 2-OH-D = 2-hydroxydesipramine 32 ng injected; 2-OH-I = 2-hydroxyimipramine, 38 ng injected; D = desipramine, 32 ng injected; I = imipramine, 36 ng injected.

## IV. RESULTS

### A. Chromatography

The separations relied on conventional, established guidelines for reversed-phase liquid chromatography (LC). The organic solvent content was increased until 2-hydroxyimipramine and 2-hydroxydesipramine were barely resolved. An octyl column was selected over a $C_{18}$ (ODS) column since capacity factors were already excessive with octyl. Using $C_{18}$ packing material would have required the use of even more organic solvent in the mobile phase, jeopardizing optimum performance from the electrochemical detector.

### B. Linearity

The liquid chromatography/electrochemistry (LCEC) system was linear for injections over the range of 10 to 65 ng of each tricyclic amine.

### C. Recovery and Detection Limits

The recovery of this method was determined by comparing the current response of spiked plasma extracts to that of a standard solution of the tricyclic amines (Figure 1). The percent recoveries calculated for each tricyclic amine are reported in Table 2. The blank plasma extracts showed no interfering substances.

Table 3 compares the minimum detectable concentrations of each tricyclic amine using EC and UV detection with a signal-to-noise ratio criterion of 5.

## Table 2
## RECOVERY OF IMPRAMINE, DESIPRAMINE, 2-HYDROXYIMIPRAMINE, AND 2-HYDROXYDESIPRAMINE FROM PLASMA[a]

| Compound | Recovery (%) | SD | CV (%) |
|---|---|---|---|
| 2-Hydroxyimipramine (64) | 68.9 | 3.7 | 5.4 |
| 2-Hydroxydesipramine (76) | 60.2 | 3.4 | 5.7 |
| Desipramine (65) | 79.9 | 5.1 | 6.4 |
| Imipramine (71) | 79.9 | 5.1 | 6.4 |

[a] Plasma concentrations in parentheses, in ng/m$\ell$ (n = 14).

## Table 3
## DETECTION LIMITS IN PLASMA, EC VS. UV DETECTION

| Compound | EC Detection (ng/m$\ell$) | UV Detection (ng/m$\ell$) |
|---|---|---|
| 2-Hydroxydesipramine | 5.6 | 18.0 |
| 2-Hydroxyimipramine | 3.8 | 18.0 |
| Desipramine | 4.4 | 28.0 |
| Imipramine | 5.4 | 17.5 |

## V. COMMENTS

During the extraction procedure, care must be taken when separating the organic and aqueous layers. A white precipitate may form at the interface of the two layers. If part of this is transferred and saved during the extraction procedure, the sample will yield a chromatogram with many interfering substances, making quantitation inaccurate.

Neither amitriphyline or nortriptyline are oxidizable; hence they will be excluded from the chromatogram. Their phenolic metabolites, however, would be detectable. Sulfate or glucuronide conjugates, if present, should be enzymatically cleaved prior to extraction if the total concentration of all forms is desirable.

## REFERENCES

1. **Hollister, L. E.,** Monitoring plasma concentrations of psychotherapeutic drugs, *Trends Pharm. Sci.*, 2, 89, 1981.
2. **Suckow, R. F. and Cooper, T. B.,** Simultaneous determination of imipramine, desipramine, and their 2-hydroxy metabolites in plasma by ion-pair reverse-phase high-performance liquid chromatography with amperometric detection, *J. Pharm. Sci.*, 70, 257, 1981.

Chapter 35

# DETERMINATION OF THIOLS BY LIQUID CHROMATOGRAPHY/ELECTROCHEMISTRY

**Laura A. Allison and Ronald E. Shoup**

## TABLE OF CONTENTS

I. Introduction ................................................................. 186

II. Principle .................................................................... 186

III. Materials and Methods ....................................................... 186
    A. Equipment ............................................................. 186
    B. Reagents .............................................................. 186
    C. Standards ............................................................. 187
    D. Procedure ............................................................. 187
    E. Calculation ........................................................... 187

IV. Results ..................................................................... 187
    A. Optimization of Chromatography ........................................ 187
    B. Linearity ............................................................. 187
    C. Interferences ......................................................... 187

V. Comments .................................................................... 187

References ...................................................................... 188

## I. INTRODUCTION

There are a number of thiol-containing compounds which are of biological interest. In living systems, glutathione, cysteine, and the corresponding disulfides are intricately involved in maintaining proper cellular function. Penicillamine, an antiarthritic agent, contains a sulfhydryl moiety, as does captopril, an antihypertensive agent. Other important thiols include N-acetylcysteine, cysteamine, and thioguanine.

Thiols are particularly well-suited for determination by liquid chromatography/electrochemistry (LCEC) using a mercury/gold amalgam. The thiols undergo a chemical interaction with the mercury electrode, which can be exploited to optimize both selectivity and sensitivity for their assay in complex biological matrices. The oxidation potential of mercury undergoes a shift in the presence of thiols, chelating agents, and halide ions, allowing measurement of these species at $+0.00$ to $+0.15$ V vs. Ag/AgCl. The specific nature of this detection reaction causes the method to be extremely selective and thus minimizes the amount of sample cleanup that is required.

## II. PRINCIPLE

In general, biological fluids containing thiols are immediately acidified. The functions of acidification are twofold: first, lowering the solution pH enhances stability of thiols to oxidation and second, protein precipitation is effected by acidification in matrices such as blood plasma. In some cases, disodium EDTA is added; EDTA serves to chelate any heavy metal ions which might catalyze oxidation of the thiols. After centrifugation and filtration, the sample is injected onto the LCEC system.

Separation of thiols is carried out using a reverse-phase column with a pH 3 monochloroacetate buffer. Sodium octyl sulfate can be added as necessary to retain the polar amino acids. The thiols are detected using a thin-layer amperometric detector with a mercury/gold amalgam working electrode at a potential of $+0.10$ to $+0.15$ V vs. Ag/AgCl.

The assay of penicillamine in plasma and urine will be discussed in detail in this chapter as a representative example of the LCEC determination of thiols.

## III. MATERIALS AND METHODS

### A. Equipment

The liquid chromatograph used for the determination of penicillamine consists of an LC-154 from Bioanalytical Systems Inc., W. Lafayette, Ind., equipped with a TL-6A Hg/Au working electrode and a Biophase® ODS 5 μ column, 250 × 4.0 mm. The liquid chromatographic system is modified to exclude oxygen[1] by replacing all teflon tubing with stainless steel. Mobile phase is continually purged with nitrogen gas, while refluxing at 40 to 50°C. Bioanalytical Systems Microfilters® are used to filter the small sample volumes through RC-58 0.2 μm membranes.

### B. Reagents

All chemicals and solvents should be reagent grade or better. For this assay, monochloroacetic acid (MCAA) was purchased from Mallinckrodt (Paris, Ky.), methanol-anhydrous reagent grade from Scientific Products, McGaw Park, Ill., disodium EDTA from Eastman Kodak (Rochester, N.Y.) and perchloric acid from Fisher Scientific (Fair Lawn, N.J.).

Mobile phase buffer is prepared by dissolving 18.8 g MCAA and 5.3 g sodium hydroxide in 2 ℓ of deionized, distilled water. Adjust to pH 3.0 with solid MCAA or sodium hydroxide.

To prepare working mobile phase, mix 100 mg sodium octyl sulfate with 80 mℓ methanol and dilute to 1ℓ with the MCAA buffer. Filter and degas prior to use.

### C. Standards

D-Penicillamine was purchased from Sigma Chemical (St. Louis). All penicillamine solutions were prepared using deionized, distilled water containing 1 g/ℓ $Na_2EDTA$ and 0.1 m$M$ ascorbic acid. Working standards were freshly prepared on the day of use, while stock solutions were stored for up to 1 week at 4°C.

### D. Procedure

Plasma: acidify 500 µℓ of fresh plasma by adding 200 µℓ of 1 $M$ $HClO_4$. Vortex briefly and let stand to precipitate proteins. Centrifuge at 1600 × g for 5 min and then filter supernatant through a Microfilter® equipped with an RC-58 membrane. Inject 100 µℓ of filtrate onto LCEC system.

Urine: acidify 2.25 mℓ of fresh urine by adding 500 µℓ of 1 $M$ $HClO_4$. Centrifuge briefly to settle any precipitate and inject 20 µℓ onto LCEC system.

The liquid chromatographic system is operated with a mobile phase flow rate of 1.5 mℓ/min. The TL-6A Hg/Au working electrode is prepared and used according to manufacturer's instructions,[1] at a potential of +0.10 V vs. Ag/AgCl.

### E. Calculation

Aqueous standard solutions or spiked plasma blanks are carried through the procedure and used to prepare a standard curve. Penicillamine concentrations are calculated by direct comparison to the standard curve.

## IV. RESULTS

### A. Optimization of Chromatography

Several assays for penicillamine determinations by LCEC using strong cation-exchange separation have been reported.[2-4] We have chosen to use a reverse-phase ion-pairing system, because of the higher efficiency and improved peak shape with available ODS columns. Mobile phase conditions were optimized by systematic variation of sodium octyl sulfate and methanol concentrations.[5] Figure 1 shows the penicillamine chromatograms obtained from spiked plasma (A) and spiked urine (B).

### B. Linearity

The detector response was linear for penicillamine over the range of 20 to 200 ng injected.

### C. Interferences

There are no known interferences with penicillamine using this method. Because of the highly selective detection process only other thiols and halide-type ions would be expected to show a response. For example, the early eluting peak in the urine chromatogram is thought to be the amino acid cysteine, based upon the retention time of a cysteine standard.

## V. COMMENTS

When working with thiol compounds, it is important to consider the stability of the compounds in both the standard solutions and samples. Thiols are easily oxidized to form the corresponding disulfides. The addition of $Na_2$ EDTA and ascorbic acid was found to suitably stabilize penicillamine in standard solutions, and sample preparations were similar

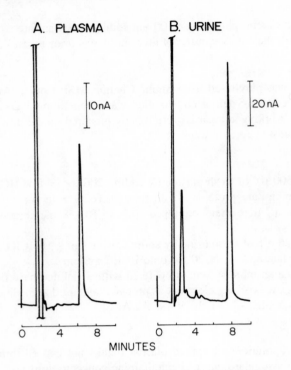

FIGURE 1. Chromatograms of (A) plasma spiked with penicillamine to 1.5 μg/mℓ and (B) urine spiked to 15 μg/mℓ. Samples were prepared as described in text and no interfering peaks were observed in nonspiked sample blanks. (Reprinted with permission of Bioanalytical Systems, Inc., West Lafayette, Ind.)

to those used in the literature.[2-4] In general, acid precipitation of proteins or acidification of urine should be effected as rapidly and reproducibly as possible to ensure stability of thiols.

## REFERENCES

1. LC-4B Manual, Bioanalytical Systems, Inc., Section 5 and 6.
2. **Saetre, R. and Rabenstein, D. L.**, Determination of penicillamine in blood and urine by high performance liquid chromatography, *Anal. Chem.*, 50, 276, 1978.
3. **Kreuzig, F. and Frank, J.**, Rapid automated determination of D-penicillamine in plasma and urine by ion-exchange high performance liquid chromatography with electrochemical detection using a gold electrode, *J. Chromatogr.*, 218, 615, 1981.
4. **Bergstrom, R. F., Kay, D. R., and Wagner, J. G.**, High-performance liquid chromatographic determination of penicillamine in whole blood, plasma and urine, *J. Chromatogr. Biomed. Appl.*, 222, 445, 1981.
5. **Allison, L. A., Keddington, J., and Shoup, R. S.**, Liquid chromatographic behavior of biological thiols and the corresponding disulfides, accepted for publication, *J. Liq. Chromatogr.*, 6, 000, 1983.
6. **Allison, L. A. and Shoup, R. S.**, Dual electrode liquid chromatography detector for thiols and disulfides, *Anal. Chem.*, 55, 8, 1983.

## GENERAL REFERENCES

1. LCEC Application Note No. 45, Optimizing LCEC Surface Chemistry with the Au/Electrode, Bioanalytical Systems, Inc., W. Lafayette, Ind.
2. LCEC Application Note No. 46, Glutathione in Whole Blood, Bioanalytical Systems, Inc., W. Lafayette, Ind.
3. LCEC Application Note No. 47, Captopril in Plasma. Bioanalytical Systems, Inc., W. Lafayette, Ind.
4. LCEC Application Note No. 48, Penicillamine in Plasma and Urine, Bioanalytical Systems, Inc., W. Lafayette, Ind.
5. **Allison, L. A.,** Series dual Hg/Au electrodes for the simultaneous determination of thiols and disulfides, *Current Separations,* (Bioanalytical Systems Inc.), 4, 38, 1982.
6. **Rabenstein, D. L. and Saetre, R.,** Analysis for gluthathione in blood by high-performance liquid chromatography, *Clin. Chem.,* 24, 1140, 1978.
7. **Perrett, D. and Drury, P. L.,** The determination of captopril in physiological fluids using high performance liquid chromatography with electrochemical detection, *J. Liq. Chromatogr.,* 5, 97, 1982.

Chapter 36

# SCREENING TOXIC DRUGS IN SERUM WITH GRADIENT LIQUID CHROMATOGRAPHY

**Pokar M. Kabra**

## TABLE OF CONTENTS

| | | |
|---|---|---|
| I. | Introduction | 192 |
| II. | Principle | 192 |
| III. | Materials and Methods | 192 |
| | A. Equipment | 192 |
| | B. Reagents | 192 |
| | C. Standards | 193 |
| | D. Procedure | 194 |
| | E. Calculations | 194 |
| IV. | Results | 194 |
| | A. Linearity | 194 |
| | B. Recovery | 194 |
| | C. Interference | 194 |
| V. | Comments | 194 |
| References | | 195 |

## I. INTRODUCTION

Rapid identification and quantitation of drugs in biofluids may be helpful in managing patients with suspected drug intoxication. Recent surveys indicate that there is an increase in the ingestion of multiple drugs.[1] Various techniques currently employed for drug screening include spectrophotometry,[2] thin layer chromatography,[3] gas liquid chromatography (GC),[4] liquid chromatography (LC),[5] enzyme multiplied immunoassay techniques,[6] radioimmunoassay,[7] and gas chromatography combined with mass spectrometry.[8] Spectrophotometric analysis is often time consuming, lacks specificity, and is prone to interference. Thin layer chromatography is a valuable technique for the detection of multiple drugs. However, it is relatively nonspecific, time consuming, and provides only semiquantitative data. Immunological assays provide high sensitivity and rapid analysis. The serious limitation of immunoassays is specificity. GC can resolve multiple drugs; however, the sample extraction methods usually employed are inadequate for resolving complex mixtures on a single GC column. To alleviate this problem, dual column GC analysis or interfaced with specific detectors, such as a mass-spectrometer have been employed. Although the use of LC has been extensively reported upon for the analysis of various classes of therapeutic drugs, there are few reports concerning the use of LC for drug screening. A simple LC screening method for the simultaneous analysis of 20 commonly abused drugs using gradient LC is presented here.[9]

## II. PRINCIPLE

Serum proteins are precipitated from 200 $\mu\ell$ of serum with acetonitrile containing hexobarbital as an internal standard. After centrifugation, an aliquot of the supernatant is injected onto a reversed-phase column and the drugs are eluted with an acetonitrile/phosphate buffer at a flow rate of 3.0 m$\ell$/min using a programmed two step gradient: the drugs are detected by their absorption at 210 nm and quantitated from either their peak heights or peak areas (Figures 1 and 2).

## III. MATERIALS AND METHODS

### A. Equipment

Liquid chromatographic systems equivalent to the following are recommended. A Model Series 3 liquid chromatograph equipped with a variable wavelength detector (LC 55 or LC 75) and a temperature controlled oven (LC 100, all from Perkin-Elmer Corp., Norwalk, Conn.), and any one of the following reversed-phase octadecylsilane columns, Waters® C18 µBondapak® 30 cm × 4 mm i.d. (Waters Associates, Inc., Milford, Mass.) or an Ultrasphere® ODS 5 µ 15 cm × 4.6 mm (Altex Scientific, Berkeley, Calif.) are used. The sample is injected into a Model 7105 valve (Rheodyne, Cotati, Calif.) mounted on the chromatograph. The column is eluted with acetonitrile/phosphate buffer at the rate of 3.0 m$\ell$/min using a two step gradient. The phosphate buffer is purged of organic impurities by passing it through a preparative column 15 cm × 10 mm dry packed with 25 to 40 µm Lichroprep® RP18 (E. Merck, Darmstadt, West Germany). This column is mounted between the pump and the mixing tee.

### B. Reagents

Acetonitrile UV grade, distilled in glass, is obtained from Burdick and Jackson Laboratories, Muskegon, Mich. The water is purified by using a multi-Q2 system (Millipore Corp., Bedford, Mass.). Phosphate buffer is prepared by dissolving 7.5 g of $NaH_2PO_4$ in 1 $\ell$ of prepurified water and adjusting to pH 3.2 with phosphoric acid.

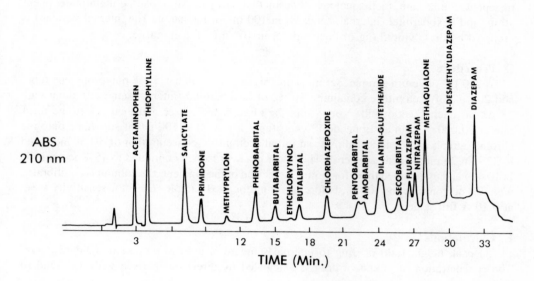

FIGURE 1. Chromatogram of a standard mixture of drugs. (From Kabra, P. M., Stafford, B. E., and Marton, L. J., *J. Anal. Toxicol.*, 5, 177, 1981. With permission.)

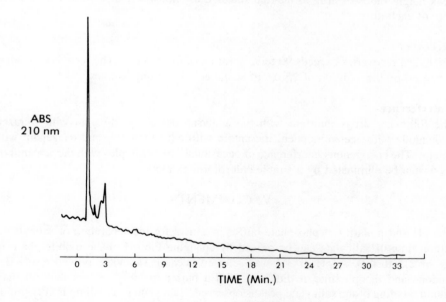

FIGURE 2. Chromatogram of an acetic acid-acetonitrile precipitation of drug-free serum. 20 µℓ of the supernatant was injected. (From Kabra, P. M., Stafford, B. E., and Marton, L. J., *J. Anal. Toxicol.*, 5, 177, 1981. With permission.)

## C. Standards

A chromatography standard mixture is prepared by dissolving 25 mg each of acetaminophen, theophylline, salicylic acid, primidone, methyprylon, phenobarbital, butalbital, butabarbital, pentobarbital, amobarbital, ethchlorvynol, chlordiazepoxide, secobarbital, nitrazepam, flurazepam, methaqualone, N-desmethyldiazepam, diazepam, glutethimide, phenytoin, and hexobarbital (internal standard) in 100 m$\ell$ of methanol. The internal standard is prepared by dissolving 5 mg of hexobarbital in 100 m$\ell$ of acetonitrile.

### D. Procedure

To 200 µ$\ell$ of serum sample, serum standard, or control in a 1.5 m$\ell$ polypropylene tube, add 200 µ$\ell$ of acetonitrile containing 10 µg of hexobarbital (internal standard) along with 25 µ$\ell$ of glacial acetic acid. Vortex-mix the mixture for 10 sec and then centrifuge for 1 min at 10,000 × g in an Eppendorf® centrifuge. Inject 30 to 100 µ$\ell$ of supernate onto the chromatograph and elute as follows: an initial acetonitrile concentration of 5% is increased to 22% in 24 min (linear gradient). The linear gradient is then continued to 45% acetonitrile in 10 min and maintained there for 5 min. At the end of the gradient, the column is equilibrated with 5% acetonitrile for 10 min before injecting the next sample. Detector sensitivity is set at 0.10 A full scale.

### E. Calculations

The peak height ratio of drug to internal standard is used to prepare a standard curve. The concentration of unknown drug is calculated by direct comparison with the standard curve.

## IV. RESULTS

### A. Linearity

Peak height ratios of drug to internal standard are linearly related from 1 to at least 100 µg/m$\ell$ of each drug.

### B. Recovery

Analytical recoveries exceeds 90 to 95% for all of these drugs. The recovery of salicylate is enhanced by the addition of 25 µ$\ell$ of acetic acid to serum samples.

### C. Interference

The following drugs interfere with the analysis: desmethyl doxepin with flurazepam, procainamide with acetaminophen, mesantoin with ethchlorvynol, and oxazepam with nitrazepam. The endogenous interference of occasional serum samples with the acetaminophen analysis can be eliminated by a simple chloroform extraction.

## V. COMMENTS

The pH and molarity of phosphate buffer is critical for the separation of salicylate. The retention time of salicylate can be moved by changing the pH of the mobile phase in the range of 2.5 to 4.0. A pH of 3.2 was selected because salicylate is well-resolved from primidone and theophylline at this pH. With a buffer molarity of less than 50 mmol/$\ell$, substantial tailing of the salicylate peak is observed. Salicylate peak tailing is very prominent on the Ultrasphere® column. We have also noticed a slight difference in selectivity between two different µ-Bondapak® columns. With both of these columns, we were able to complete the analysis in less than 35 min., however, neither one could separate phenytoin and glutethimide. On only one of these columns could we separate secobarbital from flurazepam. From these observations, we conclude that gradient conditions must be optimized for each individual column.

## REFERENCES

1. **Bailey, D. N. and Manoguerra, A. S.**, Survey of drug abuse patterns and toxicology analysis in an emergency room population, *J. Anal. Toxicol.*, 4, 199, 1980.
2. **Jatlow, P.**, Ultraviolet spectrophotometric analysis of drugs in biological fluids, *Am. J. Med. Technol.*, 39, 231, 1973.
3. **Sunshine, I.**, TLC for weak acids, neutrals, and weak bases, in *Handbook of Analytical Toxicology*, Sunshine, I., Ed., CRC Press, Boca Raton, Fla., 1975, 412.
4. **Law, N. C., Fales, H. M., and Milne, G. W. A.**, Identification of drugs taken in overdose cases, *Clin. Toxicol.*, 5, 17, 1972.
5. **Kabra, P. M., Stafford, B. E., and Marton, L. J.**, Toxicology screening, in *Liquid Chromatography in Clinical Analysis*, Kabra, P. M. and Marton, L. J., Eds., Humana Press, Clifton, N. J., 1981, 323.
6. **Bastiani, R. J., Phillips, R. C., Schneider, R. S., and Ullman, E. F.**, Homogeneous immunochemical drug assays, *Am. J. Med. Technol.*, 39, 211, 1973.
7. **Sunshine, I.**, Radioimmunoassay, in *Methodology for Analytical Toxicology*, Vol. 2, Sunshine, I. and Jatlow, P., Eds., CRC Press, Boca Raton, Fla., 1982, 205.
8. **Finkle, B. S. and Taylor, D. M.**, A GCMS reference data system for the identification of drugs of abuse, *J. Chromatogr. Sci.*, 10, 312, 1972.
9. **Kabra, P. M., Stafford, B. E., and Marton, L. J.**, Rapid method for screening toxic drugs in serum with liquid chromatography, *J. Anal. Toxicol.*, 5, 177, 1981.

Chapter 37

# SAMPLE PREPARATION FOR LIQUID CHROMATOGRAPHIC ANALYSIS

## Lane S. Yago and Thomas J. Good

### TABLE OF CONTENTS

I. Introduction ................................................................. 198

II. Liquid/Liquid Partitioning ................................................. 199
    A. Solvent Selection ..................................................... 199
    B. pH Consideration ..................................................... 199
    C. Ionic Strength Influence ............................................. 199
    D. Improving Efficiency: Multiple Extractions ........................... 200
    E. Improving Selectivity: Back-Extraction ............................... 200

III. Liquid/Solid Interactions ................................................ 200
    A. Bonded Phase Sorbents ................................................ 201
    B. Construction of the Mode Sequence .................................... 203
        1. Identify the Isolate and Its Characteristics ..................... 203
        2. Determine the Contribution of the Sample Matrix to the CMS ... 204
        3. An Empirical Approach ........................................... 204
    C. Examples of CMS Development ......................................... 204
        1. Tricyclic Antidepressants from Serum ............................ 204
        2. Benzodiazepines ................................................. 204
        3. Antineoplastic Agents ........................................... 204
        4. Nicotine ........................................................ 204

IV. Conclusions ............................................................... 207

References .................................................................... 208

## I. INTRODUCTION

Clinical laboratories are currently confronting a variety of problems deriving from the health professions' growing emphasis on drug monitoring. Increasingly, for example, newly developed drug species are being subjected to monitoring requirements that pertain not only to parent drugs but also to these drugs' metabolic products. The implications of this new focus for laboratories in both therapeutic and toxicological contexts are threefold: (1) an ever-increasing number of samples must be analyzed, (2) more complex quantitation and detection requirements must be satisfied because of the proliferation of new drugs and their associated metabolic profiles, and (3) more complex isolation and sample preparation requirements must be met as a result of the increasing number of compounds involved, many of which occur at very low levels.

Recent developments in laboratory instrumentation have focused on problems related to sensitivity (i.e., detection) as well as on problems pertaining to the large quantity of samples that must be routinely handled in the clinical laboratory. Instrumentation that is currently available offers high sensitivity with sophisticated detection modes together with microprocessor controlled, automatic equipment that allows for rapid, unattended sample processing. Regardless of the sophistication of this apparatus, however, the clinical analyst must always present samples that have been treated by a sample preparation scheme involving isolation, concentration, or both. Accordingly, this discussion will outline recent developments in chemical isolation placing special emphasis on contemporary analytical problems in the clinical laboratory.

Isolation of the compound of interest is not only the most underrated step in analytical methodology but also the most important aspect of that methodology. Unless a mode of detection is selective for the compound being analyzed, as is the case with enzyme immunoassay or radioimmunoassay techniques, the efficacy of a method's sample preparation step will ultimately determine the success of the overall analytical procedure. In order for a sample preparation method to be effective, it should be (1) *selective* in isolating the compound or compounds of interest, (2) *reproducible* and *efficient* in terms of its ability to recover the compound or compounds being isolated, (3) *rapid,* (4) *simple* with respect to actual manipulations of the sample, and (5) *cost-effective.*

Some methods require minimal sample preparation. As an example, one can add a polar organic solvent such as acetonitrile to plasma in order to precipitate proteins. The solids are separated by centrifuging the sample and then analysis occurs by injecting an aliquot of the supernatant into an HPLC system. However, a reduction in quantitation can result from potential interferences on-column; thus guard columns should be used in all such applications. Moreover, this technique does not isolate the compound from other nonproteinaceous materials that do not coprecipitate out. It has been shown that a well-developed sample preparation allows for extended HPLC column use in comparison to acetonitrile or methanol dilution methods.[1] Furthermore, dilution methods do not concentrate samples — which is a distinct disadvantage in light of increasing demands for the monitoring of low levels of new compounds and their metabolites in body fluids.

A sample preparation scheme coupled to detection proceeds in four stages. First, the sample — which can be anything from a tissue homogenate to amniotic fluid, plasma, urine, or CSF — is received and logged, then the isolation step — i.e., that stage in the sample preparation procedure in which the compound of interest is recovered — is performed. Next, the sample extract is concentrated to increase sensitivity. Finally, detection and quantitation take place by means of which data are generated on the amount of compound in the original sample.

The most important requirement in any isolation process is that the compound of interest not be denatured or destroyed. For this reason, most isolation techniques employ a physical distribution of the compound between two immiscible phases. Phase equilibria exist for all molecular species between liquid/gas, liquid/liquid, liquid/solid, and solid/solid phase systems, and these equilibria provide the basis for isolate enrichment. If the equilibrium relationship of an isolate (the compound of interest) differs from that of other unwanted compounds, chemical isolation can be achieved. Of the phase systems outlined above, the least appropriate for sample preparation criteria are liquid/gas (distillation) and solid/solid systems. The phase systems that are pertinent to clinical analysis are liquid/liquid and liquid/solid systems.

## II. LIQUID/LIQUID PARTITIONING

Liquid/liquid partitioning (i.e., the isolation of a compound through the use of the partition coefficient of that compound between two immiscible liquids) is one of the oldest and most successful isolation techniques in existence. Hence, it is also the most widely used sample preparation method. The pharmaceutical industry employs liquid/liquid extraction as a means of finally isolating manufactured drugs.

The equilibrium that exists for component I between two immiscible phases can be described by

$$[I]_{l_1} \rightleftharpoons [I]_{l_2} \tag{1}$$

where $l_1$ and $l_2$ are two immiscible liquids. For most clinical applications, $l_1$ and $l_2$ are aqueous and organic phases, respectively. The distribution of the isolate between the two phases is constant for any given temperature, pressure, pH, and ionic strength where

$$K = \frac{[I]_{l_2}}{[I]_{l_1}} \tag{2}$$

K being the partition coefficient for the given conditions. The concentrations of I in each phase are equal to the equilibrium solubilities of I in the system being considered (solvents, pH, ionic strength, etc.).[2]

On the basis of the foregoing information, it is evident that a liquid/liquid extraction is limited by the value of K. In order to maximize the partition coefficient, one must optimize several variables within the system. The key variables affecting the extraction are solvent selection, pH, and ionic strength.

### A. Solvent Selection

Clearly, not all solvents will extract a particular isolate. If for example, one evaluated the ability of different solvents to extract the bronchodilator theophylline, one would find differing degrees of recovery. Theophylline, being moderately polar, will not partition to any appreciable extent into paraffins but will partition into solvents such as ethyl acetate or chloroform-alcohol mixtures.

### B. pH Consideration

The pH of the sample may be critical for extraction. Ionization is usually suppressed so that the isolate will partition into a solvent that is immiscible with the sample.

### C. Ionic Strength Influence

Ionic strength adjustments of the sample are useful when the isolate is highly soluble in the sample matrix. Buffer salts can be used in this context; this technique often applies to the isolation of highly polar compounds from a highly polar matrix.[3]

## D. Improving Efficiency: Multiple Extractions

Inasmuch as K is a finite value and is not always sufficiently large to allow for adequate partitioning with a single extraction, multiple extractions are often employed to ensure near-complete partitioning of the isolate into the organic phase. There are many methods in which more than one extraction is required to recover the isolate. It has been shown that two aliquots of extraction solvent can recover much more of the isolate than can one total aliquot. Overall recovery and, hence, precision are affected exponentially by the number of extractions performed.[2]

## E. Improving Selectivity: Back-Extraction

By altering the pH, one can back-extract an ionizable isolate out of an original extraction solvent and into a new aqueous phase. This procedure contributes to greater selectivity, since other compounds of different pK values as well as neutral compounds will not back-extract to the same extent. The pH of this solution is then readjusted to starting conditions and the isolate is reextracted into solvent.

In summary then, liquid/liquid extractions are invariably limited by the value of K. To keep K at its maximum, one can alter the solvents, pH, and/or ionic strength.

## III. LIQUID/SOLID INTERACTIONS

In recent years, numerous sample preparation techniques using a liquid/solid approach have been introduced.[4-9] Examples of products employed for solid phases are charcoal, XAD resins (Rohm and Haas, Philadelphia, Pa.), and cartridges of bonded silica gels such as Sep-Pak® (Waters Associates, Milford, Mass.) and Bond-Elut® (Analytichem International, Harbor City, Calif.).

Why would one consider a liquid/solid approach toward sample preparation? One reason is that solid surfaces can provide a specific interaction with the compound of interest — a mechanism that differs substantially from solubility, the basis of liquid/liquid partitioning. Solids also provide large surface areas in which these interactions can take place and the condition of solvent immiscibility is not a consideration in the selection of solvents to provide interchange with the solid surface. Liquid/solid systems thus, provide a separation mechanism which is both more selective and more efficient than that associated with liquid/liquid partitioning.

Liquid/solid interactions can be described in the following manner:

$$[I]_l \rightleftharpoons [I]_s \tag{3}$$

It is true that an equilibrium relationship of the isolate can exist between the liquid and the solid surface. This relationship can be described, as in liquid/liquid systems, by defining a partition coefficient

$$K = \frac{[I]_s}{[I]_l} \tag{4}$$

If one wishes to fully exploit the advantages of liquid/solid interaction, however, one should select an appropriate solid in the presence of a given solvent system in order to force the equilibrium of Equation (3) to favor either the liquid or the solid. That is to say, one can achieve total retention by the solid when

$$[I]_l \ll [I]_s \tag{5}$$

where the partition coefficient

$$K = \frac{[I]_s}{[I]_l}$$

approaches infinity, or one can achieve total elution when

$$[I]_l \gg [I]_s \qquad (6)$$

where the partition coefficient approaches zero.

Liquid/solid interaction is not a new concept. Charcoal, alumina, and magnesium silicate, to name a few examples, are naturally occurring sorbents that have been used with varying degrees of success.[10] Although the simplicity of the approach itself may represent an improvement over liquid/liquid partitioning techniques, the aforementioned surfaces are in general nonspecific or of limited selectivity. Often, too, the surfaces are not consistent because of variability of the solid material itself. Cationic, anionic, and hydrophobic resins also have been used in liquid/solid applications for sample preparation techniques. These resins are porous, generally insoluble organic polymers whose most common form consists of a styrene-divinylbenzene copolymer in a three-dimensional structure of hydrocarbon chains terminating (in the case of the ion-exchanger resins) with ionizable groups. Those resins whose theoretical capacity for charged compounds is higher than that of any other available ion exchange materials, have been found to be useful in isolation from aqueous media of ionizable compounds such as amino acids, nucleotides, and nucleosides.

Although synthetic resins show a high capacity for ionic compounds, some difficulties are associated with their use. These materials, for example, have been found to require lengthy conditioning prior to use. In addition, because they are not rigid and cannot withstand high pressure, they must be run at relatively low flow rates. Finally, all the classical sorbents are characterized by nonspecific adsorption and thus by reduced absolute recoveries — which puts them at a disadvantage when the goal is *selective* isolation of a compound.

## A. Bonded Phase Sorbents

A bonded phase can be defined as a chemical moiety covalently bonded to a solid surface. At present, the most widely used and most effective substrate for this bonding is silica gel.

Silica gel offers many advantages over other substrates for bonded phase production. The silica gel surface provides a unique bridge so that one can covalently bond nearly any chemical moiety to the surface. If the bonding chemistry is controlled, it is unifunctional — that is, it is not confounded by other chemical groups that are initially present on the surface or generated during the bonding chemistry reaction. The bonding process can be tightly controlled to obtain a reproducible surface coverage on the substrate. Porous silica gels provide large surface areas (approximately 500 $m^2/g$), which allows for high capacities of the bonded phase moiety. Silica gel is also mechanically stable. Its rigidity allows it to withstand high-pressure differentials associated with high-velocity fluid flow, which allows sample preparation to proceed rapidly. Silica gel is also dimensionally stable in that it does not shrink or swell significantly when the solvent system is altered. Finally, bonded silica gel can withstand a wide pH range when applied in a rapid-pass sample preparation approach. The pH of the solvent will not degrade the sorbent as it will under liquid chromatographic conditions, where the sorbent is constantly bathed at high pressure with the solvent system over an extended period.

The isolating power attributed to bonded phase sorbents for sample preparation lies in the wide variety of chemical functionalities. The specific chemical interaction allows the chemist

## Table 1
### PARTIAL LISTING OF COVALENTLY BONDED PHASES TO SILICA GEL

### NON-POLAR PHASES

| PHASE | BONDED MOIETY |
|---|---|
| Methyl (C-1) | Si-$CH_3$ |
| Ethyl (C-2) | Si-$CH_2$-$CH_3$ |
| Butyl (C-4) | Si-$(CH_2)_3$-$CH_3$ |
| Hexyl (C-6) | Si-$(CH_2)_5$-$CH_3$ |
| Octyl (C-8) | Si-$(CH_2)_7$-$CH_3$ |
| Cyclohexyl (CH) | Si—⬡ |
| Phenyl (PH) | Si—⬡ |
| Diphenyl (2PH) | Si—⬡⬡ |
| Octadecyl (C18) | Si-$(CH_2)_{17}$-$CH_3$ |

### POLAR AND WEAK ION EXCHANGE PHASES

| PHASE | BONDED MOIETY |
|---|---|
| Cyanopropyl (CN) | Si-$(CH_2)_3$-CN |
| Diol (2OH) | Si-$(CH_2)_3$-O-$CH_2$-CH-$CH_2$<br>                              OH  OH |
| Aminopropyl ($NH_2$) | Si-$(CH_2)_3$-$NH_2$ |
| Primary/Secondary Amine (PSA) | Si-$CH_2)_3$-N-$(CH_2)_2$-$NH_2$<br>                      H |
| Propyl Carboxylic Acid (CBA) | Si-$(CH_2)_3$-COOH |

### STRONG ION EXCHANGE PHASES

| PHASE | BONDED MOIETY |
|---|---|
| Propyl Sulfonic Acid (SCX-P) | Si-$(CH_2)_3$-$SO_3$Na |
| Benzene Sulfonic Acid (SCX-B) | Si-$(CH_2)_2$-⬡-$SO_3$Na |
| Quaternary Amine (SAX) | Si-$(CH_2)_3$-$N(CH_3)_3$ $^+Cl^-$ |

to achieve ultraselective chemical isolation which is the overriding goal of sample preparation. Table 1 provides a partial list of currently available bonded phases.

Two terms that are commonly used in solid phase sample preparation must now be defined to better illustrate the processes that take place in liquid/solid interactions. The first of these terms, *mode*, refers to the interaction between the isolate and a sorbent in a specific stationary and mobile phase system that causes retention or elution. A mode change can be accomplished by changing either the sorbent or the mobile phase system. The second term, *chromatographic mode sequence,* or CMS, refers to a succession of interactive modes that results in the selective isolation of the compound of interest. It has been shown that each time the mode is changed, the selectivity of separation increases exponentially.[11] If one considers that efficacy of liquid/solid techniques depends entirely on the separation of molecular classes according to their chemical functionalities, it becomes clear that selectivity is considerably enhanced through the use mode changes. Generally, it is not necessary to complicate the CMS with the introduction of a great many mode changes. Because selectivity is exponentially enhanced by a single mode change, most separations need only involve a two dimensional sequence, i.e., retention on an appropriate bonded phase and elution by a solvent change. Theoretically, this two dimensional CMS should provide a 100-fold isolation factor.

## B. Construction of the Mode Sequence

The following describes the sequence of events that eventually leads to the selection of an appropriate bonded phase sorbent for selective sample preparation of the isolate out of its associated sample matrix.

*1. Identify the Isolate and Its Characteristics*

In general, the isolate should be evaluated in terms of its molecular structure, ionizable groups and associated pK values, solubility data, and chromatographic data.

**Molecular structure** — Retention by sorbents is closely related to a compound's structure. Highly aromatic compounds will be retained by different phases than will be strongly polar compounds that show a significant degree of charge density. Polarity, and resulting retention, are most often a function of the various functional groups on a molecule. Certain functionalities will have different affinities for different chemical groups. For example, an amino functionality will show an affinity for a carboxylic acid cation exchange phase. One can evaluate the opposite situation as well; if the isolate has a carboxylic acid functionality, it will show an affinity for an anion exchange phase such as a bonded amino phase.

**Ionizable groups** — Although it is often assumed that ionization should be suppressed in the performance of a liquid/liquid extraction, the wide variety of bonded phases that are currently available sometimes make this an inappropriate response. Indeed, it is often advisable to buffer a solution in order to *enhance* the ionization of a compound (i.e., a nucleotide) so that the compound can be selectively isolated on an ion exchange phase (such as an anion-exchanger). Even when hydrophobic phases such as C18 are used, the ionized isolate can be retained by the bonded phase if it possesses a large nonpolar molecular portion that is spacially removed from the ionized site.

**Solubility data** — If a compound is soluble in a nonpolar solvent, the chances are good that it will be retained by a nonpolar phase. The same logic can be applied to polar compounds as well as to polar sorbents. One should not, however, think of a bonded phase as an "immobilized solvent", as there are instances in which the rule delineated above will not be applicable. The real use of solubility data will be in the elution from the bonded phase; by choosing an appropriate solvent, one can selectively elute the isolate from a sorbent and still leave other compounds on the surface that may also have been initially retained from the sample.

**Chromatographic data** — Information on chromatographic data can be very useful in the selection of an appropriate sorbent for sample preparation. If, for instance, it is known that the compound being sought can be chromatographed on a C18 phase, this information can be translated directly back to the sample preparation procedures. As an example, if 50% organic modifer is required to elute a compound from a nonpolar chromatographic column, then 100% aqueous conditions should theoretically effect complete retention.

Once pure standards of the isolate have been tested to derive retention data on different sorbents, various solvents can be tested for their ability to elute the isolate. The solubility data gathered on the isolate is useful in this examination of elution solvents.

## 2. Determine the Contribution of the Sample Matrix to the CMS

The method as it now stands must be tested using the actual sample matrix. "Blank" matrix is useful in this context since at this point, one is not testing for the isolate but is instead testing for any interferences that may result from the methodology. Two types of interferences can arise: first, an elution interference, which could hamper detection of the isolate and second, a retention interference, which could affect initial retention of the isolate. Once one is satisfied with the results obtained, the blank sample matrix can be spiked with known quantities of the isolate and brought through the isolation scheme. A known quantity of another similar compound can then be added to the final eluant to test for absolute recovery. If absolute recovery is found to be at acceptable levels depending upon the particular recovery requirements of the individual, (typical absolute recoveries from bonded phase sorbents are between 90 and 100%), then it is acceptable to add an internal standard at the outset of the isolation scheme and to quantitate by relative recovery values.

## 3. An Empirical Approach

If enough data cannot be obtained to allow sorbent and solvent predictions to be made, an empirical approach toward developing a CMS will prove highly productive. With this method, one evaluates a spectrum of phases (nonpolar, polar, and ion-exchange) for isolate retention, and then evaluates a spectrum of solvents (polar aqueous, polar organic, and nonpolar organic solvents) for elution of the isolate. Blank sample matrix is subsequently subjected to the same CMS to which the pure isolate was subjected to test for interferences. Finally, spiked samples are evaluated and tested for absolute recovery, reproducibility, and linearity. If the empirical approach toward constructing a CMS is followed logically, retention and elution data should be rapidly developed.

## C. Examples of CMS Development

The following profiles provide examples of the sequence that one should follow in developing a CMS approach. In each case, what is ultimately presented to the HPLC is a highly purified extract of the compound or compounds of interest.

### 1. Tricyclic Antidepressants from Serum

*Solubility:* tricyclic antidepressants are soluble in nonpolar or polar organic solvents. *Chromatographic data:* it has been shown that tricyclic antidepressants are strongly retained on hydrophobic phases such as C18. This retention is so strong that C18 may not be the appropriate phase to use for sample preparation because of potential difficulty in eluting the drugs. For this reason, C2 was chosen as the phase to retain the isolates after several other phases were evaluated. *Matrix data:* the sample matrix is serum or plasma containing proteins, lipids, and potentially other drugs. In general, proteins and lipids pass freely through a hydrophobic phase on a single pass when applied under completely aqueous conditions and at a rapid flow rate, as is the case with solid phase sample preparation techniques. As described earlier, the matrix should be subjected to the CMS to ascertain if interferences are carried through the procedure. *Highlights of the CMS:* it was found that elution from the C2 phase with chloroform carried off the drugs but left many matrix components trapped on the C2 sorbent; therefore, a good isolation resulted (Figure 1). To concentrate the drugs out of the elution solvent, the chloroform was then passed through a polar cyanopropyl phase, which retained the isolates out of the solvent. This dual-phase procedure was accomplished in a single step. The isolates were then eluted from the cyanopropyl column with the HPLC mobile phase and were injected directly into the HPLC system.

### 2. Benzodiazepines

*Isolate data:* benzodiazepines are soluble in acetonitrile, alcohols, chloroform, and other

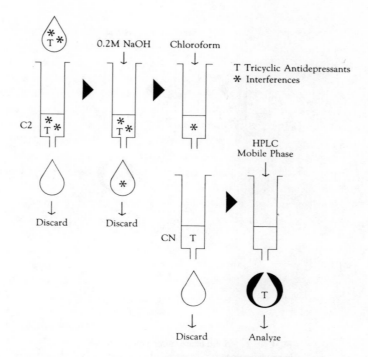

FIGURE 1. Solid phase sample preparation of tricyclic antidepressants from serum.

moderately nonpolar solvents. They can be separated on C18 phases with low percentages (35%) of organic modifiers.[8] C18 was therefore chosen to isolate these compounds out of an aqueous matrix (serum, plasma). *Matrix data:* see matrix data on tricyclic antidepressants. *Highlights of the CMS:* 50 μℓ of methanol was used to rinse the column of any remaining aqueous matrix prior to elution of the isolates with 100% methanol (Figure 2).

### 3. Antineoplastic Agents

*Isolate data:* bisantrene and mitoxantrone are soluble in alcohols, ethyl acetate, and chloroform. They can be separated on a C18 HPLC column with relatively low solvent modifier (25%). *Matrix data:* the sample matrix is serum or plasma; see tricyclic antidepressant matrix data. *Highlights of the CMS:* total elution of the isolate was achieved with a very small volume of acidic methanol (400 μℓ), since the concentrations of these compounds were low (< 10 ng/mℓ) and since high sensitivity is required (Figure 3).

### 4. Nicotine

*Isolate data:* nicotine is soluble in alcohols, chloroform, ether, and oils. High percentages of solvent modifier are required to elute nicotine from a C18 or C8 HPLC column. It is strongly retained by nonpolar phases out of alkaline solutions. *Matrix data:* urine contains salts, pigments, organic acids, and other polar molecules. *Highlights of the CMS:* since nicotine is strongly retained by C18 and C8, a C8 column could be rinsed with water to elute water soluble polar interferences (Figure 4). Chloroform was used to selectively elute the nicotine from the sorbent while leaving behind other retained compounds, such as chromophores, which did not elute with chloroform but which would have eluted if a more polar solvent (e.g., methanol) had been chosen as the elution solvent. As with the previously described tricyclic method, a polar sorbent — in this case, silica — was placed below the

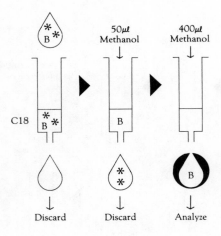

FIGURE 2. Solid phase sample preparation of benzodiazepines from serum.

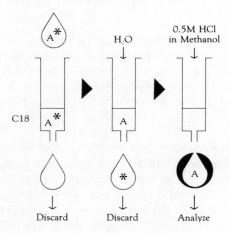

FIGURE 3. Solid phase sample preparation of antineoplastic agents from serum.

primary sorbent for the purpose of retaining the isolate out of the chloroform. The isolate on the secondary phase could then be eluted with the HPLC mobile phase. This step eliminated the need to evaporate the solvent to dryness prior to reconstitution in the HPLC mobile phase, thereby reducing sample manipulation and the potential partial loss of the

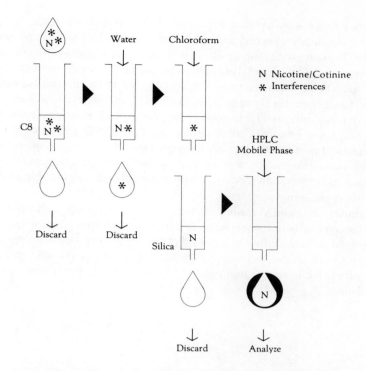

FIGURE 4. Solid phase sample preparation of nicotine and its metabolite cotinine from urine.

isolate during evaporation. This method also successfully isolated the urinary metabolite of nicotine, cotinine.

## IV. CONCLUSIONS

Chemical isolation using solid phase technology is rapidly gaining acceptance as a sample preparation technique in the clinical laboratory. Liquid/solid interactions in which the solid is a unique and controlled chemically modified surface provide the basis for ultraselective isolation of various compounds. Solid phase technology yields a rapid and simple method for sample preparation, regardless of the mode of detection that is ultimately used. The solid phase materials that are needed for sample preparation are, moreover, quite inexpensive, and the time and cost savings thus afforded result in a substantial reduction in assay cost.

Solid phase technology also provides a method for chemical isolation that can be easily automated, inasmuch as the physical manipulations, solvent requirements, and material handling steps associated with liquid/solid systems, are relatively few. Bonded phase sorbents can, in addition, be configured into devices that can be easily adapted to the instrumentation that performs the steps in a solid phase sample preparation scheme, the chromatographic mode sequence.

The application of bonded phase technology to chemical isolation adds new dimensions to the area of sample preparation. Once the isolate or isolates have been retained onto an appropriate phase, one can store the preparations or transport the samples to remote laboratories rather than having to store or ship serum, whole blood, urine, etc. Since the isolate or isolates are essentially immobilized on the solid surface, they can also undergo *in situ* reactions such as derivitizations and hydrolysis if one treats the solids with the appropriate reagents rather than performing these steps on the eluted compounds.

Bonded phase technology has now been refined to a point at which almost any chemical moiety can be covalently attached to a silica gel surface. The number of bonded phases available today is impressive, ranging from highly nonpolar phases such as octadecyl, octyl, and ethyl groups to polar phases (including ion exchange functionalities). The scope of available phases is being expanded to include special affinity functionalities, such as boronic acid moieties that permit highly specific interactions with chemical isolates. The boronic acid moiety, for example, is highly specific for the *cis*-diol functionality, which is characteristic in catecholamines and certain proteins. Other specific phases currently being investigated are chiral moeities for the separation of racemic mixtures, immobilized enzymes which achieve enzymatic activity without denaturation or consumption of the enzyme, and finally, bonded monoclonal antibodies, which are receiving considerable attention in the field of genetic engineering.

The mechanisms by which various molecular species interact with specific modified surfaces are becoming better understood in modern separation science technology. The foregoing discussion on the utility of the solid phase in sample preparation presents the state of the art with regard to this technology. The current emphasis on solid phase techniques will undoubtedly result in new techniques and products that will better achieve ultraselective chemical isolations.

## REFERENCES

1. **Kabra, P. M., Nelson, M. A., and Marton, L. J.,** Simultaneous very high-speed liquid chromatographic analysis of ethosuximide, primidone, phenobarbital, phenytoin, and carbamazepine in serum, *Clin. Chem.*, 29(3), 473, 1983.
2. **Skoog, D. A. and West, D. M.,** *Fundamentals of Analytical Chemistry, 3rd ed.*, Holt, Rinehart & Winston, New York, 1975.
3. Current, (1)3, Analytichem Technical Newsletter, Harbor City, Calif., 1982.
4. **Frei, R. W. and Brinkman, U. A. Th.,** Solid surface sample handling techniques in organic trace analysis, *TRAC*, 1(2), 45, 1981.
5. **Good, T. J.,** Applications of bonded-phase materials, *Am. Lab.*, 13(7), 36, 1981.
6. **Setchell, K. D. R.,** A rapid method for the quantitative extraction of bile acids and their conjugates from serum using commercially available reverse-phase octadecylsilane bonded silica cartridges, *Clin. Chem. Acta*, 125, 135, 1982.
7. **Zeif, M., Crane, L., and Horvath, J.,** Preparation of steriod samples by solid-phase extraction, *J. Am. Lab.*, 14(5), 120, 1982.
8. **Good, T. J. and Andrews, J. S.,** The use of bonded-phase extraction columns for rapid sample preparation of benzodiazepines and metabolites from serum for HPLC analysis, *J. Chrom. Sci.*, 19(11), 562, 1981.
9. **Canfell, C., Binder, S., and Khayam-Bashi, H.,** Quantitation of urinary normetanephrine and metanephrine by reversed-phase extraction and mass-fragmentographic analysis, *Clin. Chem.*, 28(1), 25, 1982.
10. **Perry, S. G., Amos, R., and Brewer, P. I.,** *Practical Liquid Chromatography*, Plenum Press, New York, 1973.
11. **Freeman, D.H.,** Ultraselectively through column switching and mode sequencing in liquid chromatography, *Anal. Chem.*, (53), 2, 1981.
12. **Peng, Y., Ormbrg, D., and Alberts, D.,** Improved high-performance liquid chromatography of the new antineoplastic agents bisantrene and mitoxantrone, *J. Chrom.*, 233, 235, 1982.

# INDEX

## A

Acetaminophen, 193—194
  accepted therapeutic ranges for, 4
  comparison of EC and UV detection of, 7
  liquid chromatography/electrochemistry of, 6—9
  RRT for, 177
  ultraviolet detection of, 2—4
Acetaminophen standard solution, chromatogram of, 8
Acetonitrile, 112, 144
N-Acetylcysteine, 186—187
N-Acetylprocainamide (NAPA), 40, 48—50
β-Adrenergic blocking agents, see Propranolol
Allycyclopentenyl barbituric acid, 90
Alprenolol, interference with quinidine metabolism of, 67
Amikacin
  desired therapeutic level for, 24
  optimum pH range for, 28
  precolumn fluorescence derivation of, 18—22
  spectrophotometric detection of, 26—28
ε-Aminocaproic acid
  liquid chromatographic analysis of, 12—15
  therapeutic concentration range of, 14
2-Amino-4-chloro-5-sulfamoylanthranilic acid (CSA), 112
4-Amino-4-deoxy-$N^{10}$-methypteroic acid (APA), 140—144
Aminoglycoside antibiotics, see also specific antibiotic
  precolumn fluorescence derivation of, 18—21
  retention times for, 21—22
  spectrophotometric detection of, 34—36
  therapeutic levels for, 24
Amitriptyline
  LC determination of, 174—178
  reversed-phase liquid chromatgraphy of, 168—170
  RT and RRT of, 176
  therapeutic range for, 171, 178
Amobarbital, 84, 194
Analgesics
  interference with quinidine metabolism of, 67
  morphine, 154—156
Antiarrhythmic drugs, 64—67
Anticoagulants, interference with quinidine metabolism of, 67
Anticonvulsant drugs, 72, 78—80
Antidepressants, tricyclic
  CMS for, 204, 205
  electrochemistry of, 180—183
  hydroxy metabolites of, 180
  LC determination of, 174—178
  reversed-phase liquid chromatography of, 168—171
Antidysrhythmic agents, 40—45
Antifibrinolytic agents, 12—15

Antineoplastic agents, CMS for, 205, 206
APAP, see Acetaminophen

## B

Barbiturate induced coma, 90
Barbiturates
  acetronitrile concentrations and, 87
  simultaneous analysis of, 84—85
Benzodiazepines, CMS for, 204—206
Benzothiadiazide (thiazide) diuretics, see also Thiazide diuretics, 116
Beta-adrenergic blocking agents, see Propranolol
Blood
  chlorthalidone in, 108—110
  methyclothiazide in, 120—122
Bonded phase sorbents, 201—203
Bond-Elut® extraction column, 26, 35, 164, 200
Boronic acid moiety, 208
Bromohydrochlorothiade, 116
Butabarbital, 84, 194
Butalbital, 84, 194

## C

Caffeine, RRT for, 177
Captopril, 186—187
Carbamazepine
  half-life for, 75
  metabolites of, 81
  simultaneous analysis of, 72—74
  simultaneous very high speed LC analysis of, 78—81
Cardioactive drugs, interference with quinidine metabolism of, 67
Catecholamines, 208
Chloramphenicol, 94—96
Chlordiazepoxide, 177, 194
4-Chloroacetanilide, 94
Chlorpheniramine, RRT for, 177
Chlorpromazine
  chromatograms of plasma containing, 100
  LCEC of, 98—100
  RRT for, 177
Chlorthalidone, 108—110
Chromatographic mode sequence, 202—204
CMS, see Chromatographic mode sequence
Cocaine, RRT for, 177
Codeine, RRT for, 177
Cortisol, 130—132
Cortisol/cortisone ratio, in amniotic fluid, 130
Cortisone, 130—32
Cyclopal, 78, 90
Cysteamine, 186—187
Cysteine, 186—187

## D

Deproteinization process, 144
Desipramine
  detection limits for, 183
  electrochemistry of, 180—183
  LC determination of, 174—178
  reversed-phase liquid chromatography of, 168—170
  RT and RRT of, 176
  therapeutic range for, 171, 178
N-Desisopropyl disopyramide, 40
N-Desmethyldiazepam, 102, 194
Desmethylmisonidazole, 148—150
O-Desmethylquinidine, 64—67
Dexamethasone, 130—132
Diazepam, 194
  interference with quinidine metabolism of, 67
  RRT for, 177
  ultraviolet detection of, 102—104
Dihydroquinidine, 55, 58, 65
1,7 Dimethyxanthine, 162
Diphenhydramine, RRT for, 177
Disopyramide
  chromatogram of serum containing, 44
  isocratic analysis of, 45
  ultraviolet/fluorescence detection of, 40—45
Diuretics, see also Thiazide diuretics
  interference with quinidine metabolism of, 67
  triamterene, 124—127
Doxepin
  reversed-phase liquid chromatography of, 168—170
  RRT for, 177
  therapeutic range for, 171
Drug intoxication, 84, 192—194

## E

Electron capture detection, 154
Ethchlorvynol, 85, 194
Ethosuximide
  half-life for, 75
  simultaneous analysis of, 72—74
  simultaneous very high speed analysis of, 78—81
  therapeutic range for, 75

## F

Flame ionization detection, 154
Fluornitroimidazole, 148, 149
5-Fluorocytosine, 136—138
Flurazepam, 177, 194
Furosemide
  liquid chromatography of, 112—114
  metabolites of, 112
  therapeutic concentration range for, 114
Gentamicin
  chromatogram of serum extracts of, 20, 31

  desired therapeutic level for, 24
  optimum pH for, 32
  precolumn fluorescence derivation of, 18—21
  retention times for, 23
  spectrophotometric detection, 30—31
Glucocorticoids, 130—132
Glutathione, 186—187
Glutethimide, 85, 194
Gradient analysis, 41

## H

$H_2$-blocker, interference with quinidine metabolism of, 67
Hexobarbital, 72, 194
Hydrochlorothiazide, 116—118
Hydroflumethiazide, 124
2-Hydroxydesipramine, detection limits for, 183
β-Hydroxyethyltheophylline, 160
2-Hydroxyimipramine, detection limits for, 183
7-Hydroxymethotrexate (7-OH-MTX), 140—144
Hydroxytriamterene sulfate, 124
Hypnotics
  acetonitrile concentrations and, 87
  simultaneous analysis of, 84—85
Hypolipemic drugs, interference with quinidine metabolism of, 67

## I

Imipramine
  detection limits for, 183
  electrochemistry of, 180—183
  interference with quinidine metabolism of, 67
  LC determination of, 174—178
  reversed-phase liquid chromatography of, 168—170
  RT and RRT of, 176
  therapeutic range for, 171, 178
Ion exchange phases, 202
Isocratic analysis, 41, 67
Isolate, 199

## K

Kanamycin, 18
  spectrophotometric detection of, 26—28
  therapeutic levels for, 24

## L

LCEC, see Liquid chromatography/electrochemistry
Lidocaine
  chromatogram of serum containing, 44
  isocratic analysis of, 45
  ultraviolet/fluorescence detection of, 40—45
Liquid chromatography

advantages of, 130
sample preparation for, 198
    liquid/liquid partitioning, 199—200
    liquid/solid interactions, 200—207
Liquid chromatography/electrochemistry, 6
Loxapine, 168

## M

Meperidine, RRT for, 177
Methadone, RRT for, 177
Methaqualone, 85, 194
Methotrexate (MTX; 4-amino-$N^{10}$-methylpteroylglutamic acid)
  metabolites of, 140
  simultaneous determination of, 140—144
Methyclothiazide, 120—122
5-Methylcytosine, 136
5-(4-Methylphenyl)-5-phenylhydantoin, 84, 85
Methyprylon, 85, 194
Metronidazole, 162
Microfilters, centrifuge, 6
Misonidazole, 148—150
Mode, 202
Morphine, 154—156

## N

NAPA, see $N$-Acetylprocainamide
Netilmicin
  precolumn fluorescence derivation for, 18
  therapeutic levels for, 24
Nicotine, CMS for, 205—207
Nitrazepam, 194
Nitroimidazoles, storage of, 150
2-Nitromidazole derivatives, 148—150
Nonylamine, 168
Norleucine, 12
Nortriptyline
  LC determination of, 174—178
  reversed-phase liquid chromatography of, 168—170
  RT and RRT of, 176
  therapeutic range for, 171, 178

## O

Opiates, 154—156
Oxazepam
  RRT for, 177
  ultraviolet detection of, 102—104

## P

Penicillamine, 186—187
Pentobarbital, 84, 194
  ultraviolet detection of, 90—92

Perphenazine, RRT for, 177
Phenacetin
  accepted therapeutic ranges for, 4
  RRT for, 177
  ultraviolet detection of, 2—4
Phenobarbital, 84, 194
  half-life for, 75
  simultaneous analysis of, 72—74
  simultaneous very high speed LC analysis of, 78—81
  therapeutic range for, 75
Phenothiazines, 98
Phenprocoumon, interference with quinidine metabolism of, 67
Phenylethyl malonamide, 74, 81
Phenytoin, 84, 194
  half-life for, 75
  metabolites of, 81
  RRT for, 177
  simultaneous analysis of, 72—74
  therapeutic range for, 75
Phosphoric acid, 50
Plasma
  furosemide in, 112—114
  glucocorticoids in, 130—133
  hydrochlorothiazide in, 116—118
  methotrexate in, 140—145
  morphine in, 154—156
  nitroimidazoles in, 148—151
  TCAs in, 183
  thiols in, 186—187
  triamterene in, 124—127
Prazosin, interference with quinidine metabolism of, 67
Prednisolone, 130—132
Prednisone
  liquid chromatography of, 130—132
  retention times for, 121
Primidone, 194
  half-life for, 75
  metabolites of, 74, 81
  simultaneous analysis of, 72—74
  simultaneous very high speed LC analysis of, 78—81
  therapeutic range for, 75
Procainamide
  chromatogram of serum containing, 49
  ultraviolet detection of, 48—50
  ultraviolet/fluorescence detection of, 40—45
Prochlorperazine, RRT for, 177
Pronethalol, 41, 54, 55, 58
Propoxyphene, RRT for, 177
Propranolol
  fluorescence detection of, 54—56
  interference with quinidine metabolism of, 67
  isocratic analysis of, 45
  ultraviolet fluorescence detection of, 40—45
Proprionyl-$p$-aminophenol, 6
  reversed-phase liquid chromatography of, 168—170
  RT and RRT of, 176

Psychotropics, interference with quinidine metabolism of, 67

## Q

Quinidine
  chromatogram of serum containing, 59
  fluorescence detection of, 58—60, 64—66
  isocratic analysis, of, 45
  metabolites of, 64—67
  structure of, 65
  therapeutic concentration range of, 68
  ultraviolet/fluorescence detection of, 40—45
3-OH-Quinidine, 64—67
Quinidine-10,11-dihydrodiol, structure of, 65
Quinidine-N-oxide, 64—67
2'-Quinidinone, 64—67

## R

RIA (radioimmunoassay), 130

## S

Salicyclic acid, 67, 194
Saliva, methotrexate in, 140—145
Scopolamine, RRT for, 177
Secobarbital, 84, 194
Sedatives, simultaneous analysis of, 84—87
Sep-Pak®, 200
Sisomicin, 18
  optimum pH for, 32
  spectrophotometric detection of, 30—31
  therapeutic levels for, 24

## T

TCAs, See Antidepressants
Theophylline, 194
  interference with quinidine metabolism of, 67
  ultraviolet detection of, 160—162
  very high speed LC of, 164—166
Thiazide diuretics, 112, 162
  hydrochlorothiazide, 116—118
  methyclothiazide, 120—122
Thioguanine, 186—187
Thiols, 186—188
Tobramycin
  chromatogram of serum extracts of, 20, 36
  desired therapeutic level for, 24
  optimum pH for, 37
  precolumn fluorescence derivation of, 18—21
  spectrophotometric detection of, 34—37
Triamterine
  HPLC of, 124—127
  metabolites of, 126
  therapeutic concentration range for, 126
Trifluoperazine, RRT for, 177
2,4,6-Trinitrobenzene fulfonic acid (TNBS), 26, 28, 30
Trinitrophenyl derivatives, spectrophotometric detection of, 26—28, 30—32

## U

Urine
  chlorthalidone in, 108—110
  furosemide in, 112—114
  hydrochlorothiazide in, 116—118
  methotrexate in, 140—145
  methyclothiazide in, 120—122
  nitroimidazoles in, 148—151
  thiols in, 186—187
  triamterene in, 124—127

## V

Vac-Elut® vacuum chamber, 26, 35, 78, 164